The Cult of the Wild

THE
CULT
OF THE
WILD

BOYCE RENSBERGER

Drawings by Betty Fraser

1977
ANCHOR PRESS/DOUBLEDAY   GARDEN CITY, NEW YORK

We gratefully acknowledge the New York *Times* for permission to reprint Chapter 2, "Lions," portions of which appeared in an article entitled "The King of Myths," New York *Times Magazine* of October 14, 1973, © 1973 by The New York Times Company. Reprinted by permission.

Contents

Preface

This book is as much about people as it is about animals, for it concerns the relationships between our species and those of the other animate creatures of our planet. Before man became the dominant species on the earth, it did not matter much what people thought about wild animals, beyond what they needed to know to stay clear of the predators and to kill the ones needed for food. Even so, people thought a lot about animals and invented a bewildering variety of beliefs about them.

We have inherited many of those old beliefs and, as we have become increasingly isolated from the wilderness, invented many new ones. Now that we are in direct and severe competition with the free-living animals for their once-wild territories, many of our beliefs about animals no longer serve us, or them, well. If we wish to know wild animals better, both to have a rational appreciation of wildlife and to be better able to conserve wild species, it is time to discard many of our beliefs.

This book is not intended as an exhaustive review of animal behavior. It deals chiefly with only ten major species or groups of species, each of which warrants a book by itself. Rather, it is the purpose of this volume to acquaint readers with a broader point of view about animals, a view that I believe is essential to a mature appreciation of wildlife on a shrinking planet. The ten species, or groups, were chosen because for each there is a substantial body of popular beliefs and opinion, favorable or unfavorable, that now seems invalidated by the latest knowledge gained by the zoologists who study free-living populations of the animals. In addition, there are four chapters dealing more broadly with human attitudes toward the animal kingdom as a whole.

Six of the ten animals discussed and several of the topics addressed involve Africa and it may seem that wildlife on other continents is neglected. There are three reasons for this. Africa is the last great refuge of the most abundant and diverse large-animal species. Africa has more of the larger and more popularly recognized species that have been intensively studied. And my personal experience in the subject has been predominantly with African species and problems. Nonetheless, much of what is discussed in this book applies to all parts of the earth where wild animals still live free.

For my African experience, I am deeply indebted to The Alicia Patterson Foundation which generously awarded me a fellowship to spend a full year in Africa examining both wildlife conservation and human evolution. Parts of Chapters 5 and 12 were written during that year and distributed by the Foundation through its newsletters. I am also indebted to the many excellent researchers who have studied various animal species and whose findings constitute the scientific basis for most of this book. Their names and many of their published works are mentioned in the appropriate chapters. Readers who wish to learn more about a given species than the highlights offered here are encouraged to obtain the books cited.

BOYCE RENSBERGER

1

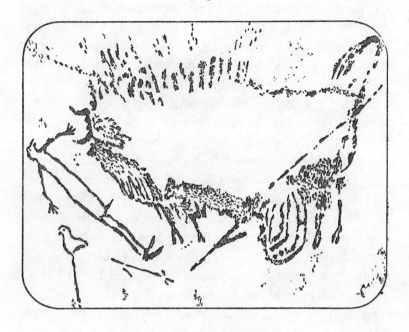

MAN and BEAST

In total ignorance of their real traits, humans have attributed to many animals a variety of human traits in the ideal or extreme. Lions are noble. Wolves are ruthless and savage. Peacocks are vain. Hyenas are slinking cowards. Songbirds are cheerful. Eagles are proud and watchful. If you want to describe a tricky or treacherous person succinctly, call him a "snake" or, to compound the myth, a "slimy snake." Few people will misunderstand. Nor is there any doubt in the listener's mind about what sort of person is being discussed if

we call him or her a "rat," a "chicken," a "vulture," a "pig."
Or, on the other side of the coin, who can help but admire
someone who is as wise as an owl or crazy like a fox, or who
has a memory like an elephant, or who is a lovable bear of
a man. No good advertising agency would let its client take
as its corporate symbol a shark (which is, in fact, an efficient
beast that has been stable over millions of years) or a sloth
(careful, methodical, precise). But many have recommended
the lion (lazy, murderous) or eagle (carrion-eating).

The list of anthropomorphized animals is long and venera-
ble. Many of the traditional traits were already well established in Aesop's day some twenty-five hundred years ago.
So widely are many of the alleged animal traits believed that,
in some cases, animal species have suffered for centuries as a
result. Wolves, for one example, have been persecuted to the
brink of extinction in Europe and the so-called Lower Forty-
eight—the conterminous United States. Hyenas, for another,
have been the victims of mass poisoning campaigns in African
national parks, particularly under colonial administrations.

Although the fashion has varied over the years, one constant in man's ideas about animals has been a pervasive if un-
witting inclination to misunderstand them.

Sometimes it is argued that animals are inferior to man,
that they are unfeeling beasts competing savagely under a
bloody "law of the jungle." The very term "animal" or
"beast," when scathingly hissed at some offending human
being, can be a powerful rebuke.

Yet at other times (and increasingly fashionable these
days) you will hear people argue that it is man who is the
savage beast, that the animals are quite humane. People contend that man is the only animal that murders his own kind,

that animals have developed a sophisticated surrender signal that mercifully ends fights before the death.

Man is the only animal that kills for reasons other than self-defense or food. How often have you heard that one? At various times in the past, that sentiment, or something akin to it, has been set forth in attempts to show how far from the "natural way" man has strayed. According to this doctrine, we have destroyed and perverted the once pure and noble ways of the natural world. The animals, it is said, are clean and pure and instinctively co-operative with each other. They do not murder or indulge in sexual perversions. Men and women, however, are alleged to have sunk far below and taken to killing others of their kind out of hate or cruelty rather than "necessity." It is argued that our destructive and wasteful life styles, especially in the industrial world, and our birth rate, especially in the underdeveloped countries, are cancers upon the earth; that while all the other animals live in perfect harmony with their environment, we are out of tune with the "law of nature."

Though our misunderstanding of wild animals has had disastrous consequences for some of the more persecuted wild species, perhaps the most unfortunate consequence is not that we have failed to know or protect the animals but that we have failed to know ourselves. We have stacked ourselves up against the myths of a pure and noble animal kingdom and found ourselves wanting. Not only is this unfair to our species, but in a time when close-up studies of wild animals can free us from the myths, it is also irrational. The wildlife behaviorists are telling us, though they seldom say it directly, that our species is not so despicable, not so hopeless. They are saying, in fact, that compared with the other species—and such comparisons seem to be popular these days—there is much to admire in *Homo sapiens* if only we can get rid of the animal myths.

In a time when the future of the human race is so much in question and when our kinship with the animal world is being so widely demonstrated, particularly through the emerging discipline of sociobiology which perceives the roots of human behavior in lower animals, it is perfectly reasonable to want to compare our behavior with that of other creatures. How much more important, then, that we have a true picture of wild animal behavior against which to measure ourselves.

Whatever our attitude toward the animal kingdom as a whole, there has always been room for a menagerie of utterly baseless views on individual species. Lions, as we will see, are revered but, as we will also see, wolves are feared. Both species have had to suffer for these extreme attitudes. Some creatures are loved (songbirds), but others are hated (rats). Some animals are considered beautiful, aesthetically valuable, and worth preserving simply for the joy of knowing they are there (deer, ocelots, giraffes), but others are thought to be ugly, offensive to the finer sensibilities, and in need of extermination (hyenas, roaches, bats).

Broadly speaking, one may categorize the popular attitudes toward animals as a kind of moral classification holding that animals are either good or bad, heroes or villains. This distinction, which has often been held even by the people charged with ensuring the conservation of wildlife, stands in the way of a mature and sensible attitude toward wildlife. One goal of this book is to discourage that distinction.

In recent years this attitude has begun to change among a very small group of people who have become interested in the conservation of all forms of wildlife for ecological reasons. These people regard all species as worth saving—not only the

lion and the hyena, but the insects as well and even the soil
bacteria so essential to the habitat. These people appreciate
the complexity and delicacy of the web that holds together
any ecosystem. Unfortunately, however, they still retain the
dichotomy of good and bad animals, with the difference that
in the "bad" category they put only one species—*Homo
sapiens*. It is a more difficult, though perhaps more important,
goal of this book to refute that distinction too.

Whatever the attitude toward various animals, pro or con,
it is remarkable how far the popular conception of supposedly
familiar species falls from the truth—at least for the small but
growing number of animals that have been studied reliably
under natural conditions.

If you wish to think of lions as noble, proud beasts, as
efficient, humane hunters, as undisputed masters of their domain, then there is no reason to try to conserve the *real* wild
lion populations in Africa or India. They might just as well all
be shot for the meat or hide. For lions aren't like that. The
real lions could conveniently vanish from the world while the
storybooks preserve the fabled beasts that many people want
to love. Only if you are prepared to accept the lion, or any
other species, on its own terms does it make sense to work for
conservation programs that will ensure a place in the natural
world for wild lions. A third goal of this book, then, is to promote the redrawing of conservation policies, or the creation of
them where none exist, in accord with the most truthful
knowledge of what various wild species are like and with an
appreciation for the fact that man is a part of nature, too.

It is usually difficult to throw out old ideas, particularly ones
that have been held for a long time, when new ones come

along unless we can examine the reasons for the origin and persistence of the old ideas.

The origins of man's relationship with the beasts must have begun perhaps three to five million years ago, when man was just starting out. The attempt to trace a modern idea back in time for more than a few centuries may seem unnecessary, audacious, or impossible. Though it may be all of these, the attempt must be made, for anything short of such a complete perspective may lead us astray. We are talking about what surely must be one of the oldest and most fundamental attitudes that has arisen in the human mind. After all, we began as animals ourselves and we remain kin, in the surest biological sense, to every other living thing. We are especially closely related to the mammals discussed in most of this book. There was a time, before the human species evolved, when we were animals by every measure. These ancestors undoubtedly included something very much like a small ape, a chimpanzeelike creature that dwelt in the forests. Now we have evolved to a point where many people find it difficult to accept that we are animals at all. Yet the molecular structure of human proteins shows them to be almost totally (more than 99 per cent) identical to the comparable proteins of chimpanzees. Since each change in a protein is the product of a corresponding mutation in the chromosomes, this means there have been very few mutations to account for the evolutionary distance between man and ape. There are fewer mutations standing between us and chimpanzees, for example, than there are between chimpanzees and monkeys. Molecular biology, then, suggests that we are, at least in our internal chemistry, still very much animals. We are a very distinctive species, of course, but animals nonetheless.

There is no agreement as to when in the course of evolution

human beings should first be regarded as distinct from their apelike ancestors. There is not even agreement on the criterion or criteria that would have shown the difference. However, since we are concerned here with attitudes, let us look mostly at the brain.

At first there can have been very little difference between man and his nonhuman ancestor. The earliest prehumans may have been something like modern chimpanzees, living in small family bands and foraging in or near the forests for edible fruits, roots, and stalks. Though they may have been a bit smarter than their ancestors, particularly in governing social relationships among themselves, they didn't show it much and were probably taken for granted by the other animals. Early man probably could wander down to a water hole and drink without disturbing the zebras or gazelles any more than would a giraffe coming to the same water hole. Herbivores (as humans once were) rarely fear one another and usually live together amiably. Territorial rights apply only to one's own species; the same piece of ground, including the water hole, can be part of half a dozen different territories, each recognized only by one species.

At this stage, man's relations with herbivores were probably quite neutral. Carnivores must have been another matter, for they were and are dangerous. The sight or sound of a leopard terrifies baboon troops and chimpanzee bands. They associate it immediately with the danger that older members of the group may well have witnessed. Early man, undoubtedly a bit smarter than the chimps, would certainly also have feared leopards, lions, and other predators. Thus our earliest attitudes toward wild animals were probably neutrality or mild curiosity toward herbivores and deathly fear of carnivores. Keep in mind that these were not merely academic positions regarding animals but emotions that were lived with every day and upon which survival often depended. In fact, a fear of carnivores may well have shaped early man's intelligence. Lacking fangs or claws and without even the other herbivores' speed, intelligence undoubtedly helped man to avoid or escape predators. Duller members of our ancestral families fell prey while the smarter ones lived to pass on the genes for their superior mentality to their descendants.

But things changed. For some reason, which remains unclear, man's ancestors did not remain herbivores. They took to eating meat as well, and that inaugurated the first of three great revolutions or transformations in relations between man and the other animals. The taste for meat may have developed when opportunistic vegetarians sampled a dead carcass, perhaps an abandoned lion or hyena kill. Those who had the teeth for it and digestive systems that could handle animal protein thrived. Their improved nutrition made them stronger. The risk of mental retardation due to protein deficiency in a drought was reduced because long after the plants dried up, prey would survive, however weakened. And the big brain accelerated on its course toward us. Relations with other animals would never be the same again.

Early man's attitudes toward the carnivores probably remained much as they had been, but there may have been the beginning of an admiration or a respect for carnivores' hunting abilities. Observation of a pride of lions working cooperatively to make a kill, for example, may have given early man a lesson in improved hunting techniques. Man's views of the herbivores, however, must have changed dramatically, as did herbivores' views of man. They were now quarry and they learned to fear man.

"With the origin of human hunting, the peaceful relationship was destroyed and [ever since] man has been the enemy of even the largest mammals," anthropologists Sherwood Washburn and C. S. Lancaster, both of the University of California, Berkeley, wrote for a conference on man's hunting past some years ago. "In this way the whole human view of what is normal and natural in the relation of man to animals is a product of hunting and the world of flight and fear is the result of the efficiency of the hunters."

This state of affairs lasted from the beginning of early man's appearance three to five million years ago and probably did not begin to change significantly until about fifteen or twenty thousand years ago with the beginning of agriculture. Man's most deeply ingrained attitudes toward animals, then, are those of hunters—not sport hunters but people who had to hunt to live. Though they were close to nature, our hunting ancestors were actually not in as close contact with animals as we might think. They avoided predators for obvious reasons, and herbivores, for equally obvious reasons, avoided man. Although they obviously came to know the comings and goings of various animals—enough to avoid some and find others— the rest of their animal lore was probably based as much on fantasy and superstition as on fact.

Washburn and Lancaster cited an incident in modern times when a member of the Hadza tribe, which still lives in Tanzania by hunting, was taken in an automobile to Nairobi National Park, just outside Kenya's capital city. The man was amazed and excited by the abundance of animals that grazed peaceably nearby as they drove about. Though he made his living as a hunter, in his entire life he had never seen so many animals at once. The animals in the park, protected from human hunters for many years, reacted to people in cars as if they were other herbivores.

Washburn and Lancaster did not say anything more about the Hadza hunter's feelings during that visit to the game park, but it is a good guess that the hunter probably wished he had his hunting weapons with him. That is how sport hunters often feel when they visit one of the African game parks. A few shots, however, would soon remind the animals what sort of creature man really is, and the more "normal" relationship between predator and prey would be quickly restored.

The Hadza hunter's reaction was probably not unlike that of early man when he began to develop his hunting prowess. There is archaeological evidence in Africa, Europe, Asia, and North America that as big game hunters, as this way of life is known among anthropologists, entered virgin territory, they killed as many animals as they could. The fossil records in these areas reveal abrupt extinctions of many big mammal species at about the times the hunters arrived. The most amply documented evidence of this comes from North America, where efficient big game hunting did not appear, according to the most widely held expert opinions, until about thirteen thousand years ago.

Paul S. Martin, a University of Arizona anthropologist, has examined the fossil remains in North America of such giant mammals as the mastodon, mammoth, and giant ground sloth and found that all became extinct at about that time. He has offered evidence that a wave of extinction radiated from the region of Alaska southward and eastward through North America. Although man was in the Western Hemisphere long before this time, the evidence suggests he had depended on a broad spectrum of plant and animal resources and lacked hunting weapons and skills to kill the larger mammals in significant numbers. Around thirteen thousand years ago, however, either a new set of weapons was developed or a new wave of human migration across the then-exposed Bering land bridge brought a new way of life to North America. The people of those times were identical to ourselves and possessed as much intelligence as we do. In a relatively brief period, Martin argues, this new tradition of almost exclusive subsistence by big game hunting exterminated nearly all the largest mammals.

Behind the advancing front of big game hunters, however, some people remained, to prey on the smaller animals. After the first rampage of slaughter, they changed their hunting ways of necessity to a more measured pace that would allow a certain portion of animals to survive. Smaller animals are more numerous than big ones to begin with, but still, wholesale slaughter could not be allowed to continue if there were to be any animals left to reproduce for the following year's harvest.

Quite likely this same pattern of massive overkill followed by sustained-yield hunting took place all over the world, wherever man went and whenever a new hunting tool—say, a sharper spear point or the newly invented bow and arrow— increased man's hunting efficiency. A similar pattern, incidentally, is being re-enacted today on the oceans where modern man is carrying out another rampage of wholesale slaughter, this time against whales. With each new advance in whaling technology, there has been an upsurge in the rate of killing. Japanese and Soviet whalers, using sophisticated whale-finding electronics and huge factory ships, have in recent years been killing whales at such a pace that many wild populations are now seriously declining and conservationist pressures are mounting for a more modest rate of killing that will allow a residual whale population to survive.

As unbridled hunting became increasingly improvident around the world, early man had to tame his initial tendencies and increase his efficiency. One problem with killing a large animal is that most of the meal spoils before it can be eaten. Even a relatively small carcass—for example, that of a deer—provides more meat than the typical hunter's family can eat before it goes bad. Thus it may well have been that those hunters clever enough to devise preservation and storage methods, such as drying or salting or burial in cool earth, had more protein over a sustained period of time than did their duller cousins. Again, intelligence and an ability to plan ahead would prevail.

As the abilities of early man grew, he came to look at the animals of the forests and plains not only as food on the hoof but also as living beings something like himself. Animals, he could see, were often anatomically organized much like people—eyes, nose, mouth, and ears positioned roughly the same way on a head attached by a neck to a trunk with four limbs. Like people, some animals hunted their food while others foraged it. They sometimes lived in families, mothers suckling the young. Usually families were peaceable, but sometimes there were quarrels. When danger threatened, the bigger ones usually defended the smaller ones. In other words, there were enough obvious similarities between man and the animals that it no doubt was as easy then as it is now to project some distinctly human, or anthropomorphic, traits onto animals. Slow and awkward appearing animals must be stupid; fleet and agile creatures must be quick-witted. Nocturnal animals must be evil and up to no good, while those that are abroad in daylight must be pure and open.

How, then, our ancestors may have wondered, did the animals regard people who went about killing animals much of the time? Projecting human values, our ancestors may have reasoned that the animals would respond the way people do when members of their clan are killed.

Richard Lewinsohn, a sociologist, has theorized that out of this tendency to project human ways onto animals came one of primitive man's first set of artificial ideas about animals. "Mutual help and revenge are most intimately bound up in the human spirit," he wrote in his book *Animals, Men and Myths* (1954). "Must it not be the same with animals? And will they not take revenge if man kills some of them, even as men of the same clan take revenge if one of them is killed or carried off into captivity? Fear of animal revenge is the great leitmotiv resounding in all the social and cultural prescripts lumped under totem and tabu."

If hunters slew an elephant, might the beast's surviving kin not come to seek revenge? Or if they killed a wildebeest, prize quarry of the lion, might the lions not try to punish the men for raiding their herds? Somehow, our ancestors may have reasoned, ways must be found to placate or appease the offended animals. Since man needed meat to survive (vegetables in a preagricultural society would rarely have provided adequate nutrition), some kind of compromise must have been worked out to balance man's fear of revenge.

One method must have been to make animals objects of veneration and homage. Idolize and idealize animals, our ancestors must have thought, and that will atone for the sin of killing animals. Thus we have, in the oldest known works of man not made for a practical purpose, images of animals in all their magnificence. In the cave of Lascaux in France and in many others around Europe and on rock faces in North and East Africa are frescoes and incised drawings of extraordinary beauty and style. There are animals with great swooping horns, running animals, mating animals, slain animals, magnificent animals, fearsome animals. In one scene at Lascaux there is a herd of horses within a frame of deer heads, together with cows, bulls, bears, lions, a rhinoceros, and a buffalo which has just killed a man with a bird's head.

Although the real meaning and purpose of cave art has long been controversial, there is little doubt that it was meant to serve some magical or religious purpose. Many of the scenes are in remote and cramped parts of caves where no one but the artist would be likely to go. When the artist had paid his tribute, the only witness to his devotion would be Mother Earth herself.

Veneration of animals has, of course, been a staple of human cultures ever since. Ancient Egypt had dozens of animal cults. Many African and American Indian tribes made certain species their totems, or mascots, and, by venerating and protecting living members of that species, hoped to gain its powers and its protection.

As Lewinsohn observed, people have always looked at animals and seen abilities and powers they lacked. To fly like a bird or run like a deer or to be as strong as a bear or a lion have been wishes of human beings everywhere, from jungles to zoos. "Jealousy and the lust for power gnaw at him," Lewinsohn wrote. "Man wants to appropriate the animal qualities he lacks. He wants to add them to his own to increase his power."

But how? Lewinsohn has theorized one possible scenario: A group of hunters sits around the campfire at night. One, having thought of a way to have a bit of fun, sneaks off and gets the head and skin of the beast they have killed that day and just consumed. Disguised as a bear or whatever, the man creeps up on his companions, grunts, and gives them all a fright before laughing at how he has tricked them. From such playfulness, Lewinsohn hypothesizes, may have come the idea that a hunter could simply disguise himself to sneak up on animals, increasing the chance of making a kill.

"When the game was repeated and also resulted in a surprisingly large bag," Lewinsohn continues, "the disguise became an emblem, a mascot, believed in as anything is believed in when it ensures success. As other members of the clan followed suit, the animal head developed into a collective symbol. Thus the luck-bringing animal became the totem of the clan." It is also possible that the idea of using an animal-skin disguise simply came from some clever hunter who figured out by himself that it might fool living animals.

The phenomenon of venerating of animals as a balance or atonement for killing animals persists today in a remarkably similar form among sport hunters and conservationists. The sport hunter's trophy room, with its mounted heads and boastful photographs, is not such a far cry from the animal carvings, drawings, and sacred skins and skulls venerated by primitive hunters. Many of the activities of lay conservationists can be similarly interpreted. Take, for example, the enormous trade in wildlife photographs, picture books, sculptures, and other animal-oriented products that is operated by the various wildlife conservation organizations. Many such groups have turned the animal-appeasement tradition into a direct way of benefiting wildlife by channeling funds raised from the sale of such materials into conservation and wildlife research programs. These efforts are entirely laudable and worthwhile, but is not the practice quite similar to the sale of religious objects and relics as aids to worship and do not people indulge in such practices out of some sense that they are demonstrating their faith and devotion?

Thus, in at least one interpretation of modern animal-related behavior, it is possible to see activities and motivations not too different from those of our hunter ancestors, who wished to atone for their killing. The killing of wild animals continues today, of course, and those who most deplore it are the educated citizens of the industrialized world who, in many cases, are at least indirectly responsible for the destruction of much wildlife, as will be explained in the last chapter. It is from these same comparatively wealthy people that most of the animal veneration and sacrificing (usually financial) to the animal kingdom comes today. Can it be that an ancient traditional attitude toward animals persists today in those who fear that if we meddle too much in the animal kingdom, some dire ecological catastrophe will befall mankind? Although it is true that we possess scientific knowledge suggesting that some catastrophes would follow some forms of meddling, the overall risk of catastrophe is nowhere nearly commensurate with the widespread popular fear that no extinction must ever be allowed to take place again. (This point is developed further in Chapter 14.)

Whatever attitudes toward animals we may have brought with us from our hunting past, they are not the only ones we now have. For at some time in the past—it varies from place to place—hunting began rapidly to fade as mankind's chief means of livelihood. Agriculture replaced it and man's attitudes toward wild animals underwent the second major transformation. Wild animals of all kinds became man's direct enemies.

The crops and domesticated animals that farmers kept were constantly threatened by predators and herbivores. The lion, always a threat to people, now also became a menace to the flocks of sheep and herds of goats kept by early pastoralists. The wolf, never much of a threat to man despite popular impressions and, in fact, from whose amiable ranks the dog was domesticated, now became hated for its predation upon the tame herds. More dramatically, those smaller plant-eating animals that had not been major food sources for people and to which people were largely indifferent, became man's enemies when they chose to forage among the crops. Even the birds and insects now became enemies because of their attacks on crops. In short, for the bountifulness of agriculture, which enabled human populations to surge in numbers and to establish settled, civilized villages, man had to declare war on wildlife —all wildlife; the only good animals at that time were the domesticated ones. An early agriculturalist's idea of paradise, which can be deduced from the vision of Eden given in Genesis, gives man dominion over all the beasts of the planet.

Earlier, as hunters and gatherers, people had to live in close harmony with nature. Although they probably did not know the details of animal behavior, at least they had to know enough to fit their own lives around those of the animals. They had to know the comings and goings of animals and the life cycles of edible wild plants. They moved out from a simple home base and spent much of the day roaming about among the animals in search of food. The land could in no sense belong to them, for it was the province also of the beasts to whom homage and respect, perhaps even worship, were due. The farmer, on the other hand, is tied to one place. His fields and flocks must always be tended and defended against the wild beasts. Though the farmer attunes his life to the seasons and although many farmers cherish a feeling of harmony with nature, the fact is that a farmer's success depends on his winning battles against pests and predators and even against nature's efforts to recolonize his fields with "weeds." The hunter, by contrast, succeeds by changing little or nothing in nature. He slips quietly through forests and plains trying to attract no attention, often even covering his tracks lest they betray his presence to predator or prey. Modern-day farmers in industrialized countries are in a somewhat different position in relation to the natural world than were their primitive predecessors. Whereas the early cultivators carved their fields out of wilderness, today's Western farmers are replowing land that was tamed decades or centuries ago. Though they must still war with insect pests and "weeds," they share some of the renewed interest in wildlife conservation and have adopted many practices that favor the return of some forms of wildlife such as birds.

As the coming of agriculture changed man's ideas about wild animals, it also changed wild animals. As the human population expanded, it required that more and more land be claimed from the wilderness and turned into pasture or cropland. Wild habitat was destroyed. Like ripples spreading out from the population centers, the farmland pushed away the wild animals until, even thousands of years ago, it was not uncommon in many areas for people to spend their lives miles away from the wilderness and its animals.

On the fringes of the settled areas, of course, wild animals could still raid crops or carry off children but all that most people knew of these events was hearsay and, inevitably, exaggerations of reality. As agriculture moved from an innovation to a tradition in any given place, a process of alienation from wild animals began, and pushed man's attitudes toward the beasts further into the realm of fantasy and conjecture.

The third great transformation in man's views toward animals came as villages grew into cities, where large numbers of people, particularly among the educated (privileged) classes, had almost no real contact with wild animals and no genuine stimulus to think of a given species, if at all, as a friend or an enemy. Consequently, the way was open to invest any animal species with any supposed attribute, regardless of how well it fit reality. For the urban peoples of that time, and for many urbanites ever since, wild animals had no reality beyond whatever learned men—themselves isolated urbanites—said.

The villages quickly grew into cities, with specialized forms of labor and commerce, that are the direct forerunners of our own. The great civilizations of Mesopotamia, Egypt, China, and the Aegean arose during those times, between four thousand and ten thousand years ago. All the original civilizations that are known were rich in animal beliefs and poor in animal knowledge. Perhaps the most famous vehicle in Western culture for those ancient animal ideas was Aesop, a legendary Greek slave who, some twenty-five hundred years ago, is thought to have collected or invented fables using animals as characters. Little is known of Aesop. Aesop was said to have been born a slave, but his wit and storytelling ability brought him to the attention of King Croesus, who invited Aesop to his court in Lydia where his fables gained further renown.

There is no reason to suppose that Aesop or any of the other fable makers were attempting to portray animals realistically or that they had much reliable information on which to do so. The use of animals in the fables was simply a literary device for holding the listener's interest. The object of a fable is, after all, to get a person to look at human ways of behaving and thinking in a new light. The use of talking animals permits the listener or reader to analyze the foolishness or wisdom of some animal behavior from a distance and only then to realize how human it is. The device itself, the talking animal that acts like a person, has never failed to amuse.

Through repetition of fables, ancient peoples came to associate certain human traits with certain animals. Isolated from wildlife and having no intrinsic reason to like or dislike a wild species, civilized people could perceive almost any animal as having almost any human trait, good or bad. These ancient moral classifications of the animal kingdom have largely persisted to this day.

Increasing alienation from the world of wild animals characterized the inhabitants of Western Europe from the dawn of civilization until quite recently. During the medieval period, moral classifications were very much in vogue and as far as animals were concerned, things got little better in the ensuing periods of the Renaissance and intellectual enlightenment. In fact, it was in the times of supposed rationalism that Western society saw the beginnings of the most destructive myth of all —the beastliness of man. Instead of starting from scratch to evolve a sensible view of man's place in the natural world, philosophers like Montaigne, in the sixteenth century, simply stood the old order on its head and claimed that animals were morally superior to human beings.

"We give ourselves superior rank over other creatures out of foolish pride and stubbornness, rather than because of solid justification," Montaigne said. Although it was a genuinely bold stroke to argue that man was a member of the animal kingdom, as he did, Montaigne went even further. Man's baseness, as he saw it, was the result of a moral sensibility that was not as highly developed as that in many animals. The animals, he said, have a true sense of brotherhood and justice; they have languages of their own and intellects occasionally surpassing man's. Man's chief distinguishing attribute, Montaigne argued, was his power to create new thoughts, to dream, to imagine.

The real nature of animals was not a subject for idle speculation in those times. It was the meat and blood of many a
heated philosophical debate. In the seventeenth century Des-
cartes, for example, passionately maintained that animals
were nothing like people because they couldn't talk. His student Spinoza agreed partly but argued that animals did have
feelings. Leibniz, a few years later, went further than Montaigne, and said animals had souls and that the only difference
between theirs and human souls was in the quantity, not the
quality, of their souls.

For all the reality in the debate, the philosophers might as
well have been arguing about how many unicorns could dance
on the head of a pin. The philosophers argued that man's arrogance toward the animal kingdom was not justified, and
they laid down a philosophical basis for the view that animals
have rights and claims on man as valid as man's rights and
claims on animals. Although this emerged from a perfectly
reasonable supposition that we are as much a part of nature
as the animals, some of the philosophers got a bit maudlin
and acted as if reparations were due the animal kingdom. Sen-
timentalist thinking held sway to the point that in the nineteenth century Schopenhauer, in his will, left a huge sum for
the care of his dog and Nietzsche once walked up to a horse
in a Turin street and, as if it were a long lost kinsman, em-
braced it.

Although the loss of arrogance was a step forward, things
went too far, for, as Montaigne had held, in the moral realm
human beings were seen as inferior to the noble lion, the
steadfast elephant, and even the precise and orderly tuna,
which Montaigne assumed to have a high mathematical sense
because they spaced themselves within the school in certain
geometric patterns.

One of the best known celebrants of the purity and intrinsic worth of animals was William Wordsworth whose poetry of the late eighteenth and first half of the nineteenth centuries is considered pivotal in the transition, then taking place, from the Age of Reason to the Age of Romanticism. Wordsworth wanted to turn the thoughts of his readers from what he considered to be the sordid affairs of technological man to the pristine glories of unsullied nature. Man, Wordsworth believed, had turned away from nature and lost his soul. Wordsworth said man entered the world "trailing clouds of glory" but drifted into a baser existence. Although much of Wordsworth's romanticism is given over to a perfectly reasonable, tranquil appreciation of the outdoors and the ability of wilderness to soothe and relax Westerners, some of his writings, if taken literally, now seem ludicrous. In his famous *Ode on Intimations of Immortality* Wordsworth describes the glories of the natural world:

> And all the earth is gay;
>   Land and sea
> Give themselves up to jollity,
>   And with the heart of May
> Doth every beast keep holiday; . . .

> Ye blessèd Creatures, I have heard the call
> Ye to each other make; I see
> The heavens laugh with you in your jubilee;
>   My heart is at your festival,
>   My head hath its coronal,
> The fullness of your bliss, I feel—I feel it all.

Whatever Wordsworth felt, it was not, of course, the good "vibes" of ecstatic songbirds or frolicking hosts of golden daffodils. More likely Wordsworth was speaking out of his revulsion for what he considered to be the cruel work of man. In a brief poem entitled *Lines Written in Early Spring,* this view seems evident. The final four stanzas read as follows:

> *Through primrose tufts, in that sweet bower,*
> *The periwinkle trail'd its wreaths;*
> *And 'tis my faith that every flower*
> *Enjoys the air it breathes.*
>
> *The birds around me hopp'd and play'd,*
> *Their thoughts I cannot measure—*
> *But the least motion which they made*
> *It seemed a thrill of pleasure.*
>
> *The budding twigs spread out their fan*
> *To catch the breezy air;*
> *And I must think, do all I can,*
> *That there was pleasure there.*
>
> *If this belief from heaven be sent,*
> *If such be Nature's holy plan,*
> *Have I not reason to lament*
> *What Man has made of Man?*

And if these beliefs were not sent from heaven and if they are not even part of nature's plan, holy or not, what then? Wordsworth doesn't contemplate that alternative. There was, apparently, no question in his mind that nature was exactly what he thought it was. Any serious effort to determine objectively how the wild creatures really behaved would not begin for at least a hundred years after Wordsworth's death. For much of the time in between, the appealing, storybook image of the peaceable animal kingdom pervaded much of popular thought on wildlife.

Although natural science took great leaps forward during

last half of the nineteenth century, especially with Charles Darwin's epochal work, the man-is-beastly school of opinion continued, for almost none of the scientific advances had to do with animal *behavior*. Most of the things scientists studied in the wild animal kingdom, still becoming ever more remote because of the steady expansion of people and their farms, were dried or pickled specimens sent home by voyaging naturalists. The occasional caged, live specimen behaved very little like its free-living relatives. Even the accounts of explorers who observed a wild species in their native habitats were of little genuine help. How many accounts, for example, of a respected, fearless explorer facing chest-beating, long-fanged gorillas would one have to read before believing these gentle apes to be dangerous?

During the late nineteenth and early twentieth centuries another factor played a significant role in further distorting Westerners' ideas about wild animals. This was the rise of the big game sport hunter, particularly the variety that sought exotic species far from home. As long as hunters were taking species from close to home, they did not have much of a story to tell the nonhunters except the truth. But with increasing affluence, more and more sport hunters could travel to distant places and shoot creatures that were unfamiliar to the folks back home. One could hardly hope to impress people with one's exploits or manhood by bagging a turkey. But bring home an elephant's tusks or have the taxidermist set the bear's jaws into a perpetual snarl and most people will believe any story you care to tell about the hunt. Never mind that elephants are loathe to attack an intruder and usually walk away when approached; never mind that they are almost as big as the side of a barn. Few people who look at those bear fangs and claws will know they are for ripping open rotten logs to get at the termites; nor will they know that when a bear bares its teeth, it's not preparing to attack but is terrified and hoping that you will be frightened away by a show of teeth.

Sport hunting, particularly of exotic species, is done not only for the adventure of the moment, but, quite plainly, as a proof of one's prowess both to one's self and to others. And very much a part of this ego-bolstering process is fostering belief that the hunt is dangerous. The most prestigious African game animals, in the eyes of hunters, are the so-called Big Five, alleged to be the most dangerous game. They are, the lion, leopard, rhinoceros, Cape buffalo, and elephant. A hunter who bags each of the five has achieved huntingdom's quintuple crown and could, at least until recently, gain the respect of most any man he cared to brag to. Even some women are impressed.

But the big game hunter's principal legacy, aside from rows of dusty, disembodied heads and the depletion of many species all over the world, is another contribution to the alienation of urbanized people from wild animals. Their tales have added to the illusion that the wilderness is a dangerous place where only the brave go and then usually accompanied by a retinue of gun bearers. This belief is still surprisingly strong in urban peoples today who find it a bit scary to spend a night in the woods. Can it be that we still retain the primal fear that all the good animals are asleep at night and that only bad ones are active then?

The long process of alienation of Western and industrialized man from wildlife has today gone about as far as it can. The assortment of opinions about any unfamiliar species would seem to be essentially random. Generations of fables, confining zoos, and fading instincts have put us at such a distance from the beasts that, in our efforts to recapture an elusive sense of "oneness with nature" we assume that that state must be some kind of utopian paradise—clean, pristine, unspoiled. We have lost touch with so many animals that we are unable to have a realistic concept of nature as a whole. We assume that "unspoiled" nature, meaning untouched by human hands, is what is merchandised in many of the conservation organizations' magazines and picture books. The sun always shines on glistening dewdrops. Lordly elks graze in meadows of brightly colored wild flowers stretching to a stately forest in the background. Sparkling streams splash over craggy rocks as a glossy black bear carries off a fresh-caught fish. There is rarely a sign of human presence, for that might spoil the scene.

How we long to enter such a world and recapture lost innocence. While the photographs are almost always of genuine natural settings, they are very carefully selected to cater to the belief in a natural world that is pristine only when there aren't people in it or affecting it. If it weren't for people, the subliminal message says, the whole world would be like this. Man is the spoiler. Man is the villian. It is all black and white. Nature is good but man, somehow not part of nature, is bad.

It is the measure of our alienation from wildlife that we conceive of it as something apart from ourselves. The most pathetic aspect of this alienation is that when we do try to appreciate wildlife, we do it with the blinders of centuries of myth and misunderstanding.

We marvel over the photographed wild flowers in the nature magazine and ignore the insects prowling in our hedge. We stare at the sleepy lion in the zoo and struggle to make it match an image of regal serenity. We watch a white-tailed deer bound across the road and think of Bambi.

It is time to put away this impoverished, alienated view of nature. It is time to recognize that man is also a part of nature and that there really never was any such thing as a constant or untouched wilderness. The glory of the natural world is not that everything was created just so. It is that the earth has always seen shifting natural panoramas, with one species rising up, spreading out, and changing the face of the earth while others recede and even die out. It is a drama that is constantly unfolding, and however much we think we deserve only to be in the audience, we are on the stage.

2

LIONS

Whether reposing serenely in marble before some public building or roaring from the screen at the start of an M-G-M epic or supporting the crest in a royal emblem, the lion has been for centuries the world's undisputed symbol of nobility and majesty.

King of Beasts. Lord of his domain.

Don't believe it. Most of what the world thinks it knows about lions was learned from poets, pet keepers, and press agents. It is not true, for example, that lions kill only for food, or that they are skillful hunters dispatching their victims with merciful swiftness, or that mother lions protect their cubs from all harm, or that lions are above scavenging rotting carcasses. Recent studies of lions living naturally in Africa—the first scientifically solid studies of lion behavior—prove that these and a host of other common notions about lions are false. Although there is no scientific measure of an animal's more subjective attributes, the findings hardly support the notions that lions are noble or courageous or even especially lovable. What the studies *do* show—and what a brief look at the findings will demonstrate—is that while lions do not live by the best of human value systems, they do behave like . . . well, like lions.

All this is not to say that they are not worth preserving as part of the earth's wild heritage. Far from it. But part of a mature appreciation of wildlife should be a desire to understand each animal species as it truly is. The lion mania of the last few years suggests that we are still in a more adolescent stage of wildlife appreciation. Lions today are probably the most popular, the most respected and revered, and perhaps even the most loved of wild animals. They are also the most misunderstood.

As if the old notions of lions weren't good enough, the image of *Panthera leo* has been embellished—Disneyfied, if you will—by a series of calculatedly heartwarming books and movies about Elsa, the lovable lioness, raised in East Africa from an orphaned cub by George and Joy Adamson. Millions thrilled when they read or saw that, true to an animal story much older than *Bambi,* the grown-up Elsa went into the wild to find a mate and returned to show her cubs to the kindly human beings who were her foster parents.

Although *Born Free,* by the Austrian-born Mrs. Adamson,

was published in 1960, the book and a couple of spin-off books, *Living Free* (1961) and *Forever Free* (1963), are still in print and available in many bookstores. The Elsa cult is today bigger than ever. Tens of thousands of children belong to "Elsa Clubs" all over the world. The Elsa Wild Animal Appeal raises millions of dollars to aid conservation projects in several countries.

Spurred by the Elsa phenomenon and a growing interest in wildlife in general, entrepreneurs across North America, Europe, and Australia have opened up scores of "lion country" parks where people can drive their cars among "wild" lions gnawing on haunches of institutional grade beef. More than two dozen are now operating in the United States alone. Many of the parks record over a million customers a year.

In none of the commercial parks, of course, do the lions live any more freely than barnyard animals. In most, if not all, the beasts are rounded up like cattle each evening and herded into small huts or cages for the night. For a mainly nocturnal species like the lion, this is the cruelest time to shut them away. It would be more natural if the lions were caged during the day, when they prefer to sleep, and released at night when they are naturally more active. But then, a lion's principal activity is hunting, and no "safari park" is likely to let the lions do that. After all, children come to the parks. How would parents explain if they found a lovable lion ripping into the belly of a cute zebra, lapping up its entrails? (In the wild, lions nearly always seem to start with the entrails, apparently considering them a delicacy.) And, what's more, it would cost the parks more to replace the zebra than to buy another lion. Unlike zebras, lions breed readily in captivity and, as more and more parks breed them, a glut of lions is developing that is forcing the price still lower. Some parks actually give their lions contraceptive drugs to avoid the problem of having cubs they don't need and can't dispose of.

In East Africa lions were largely considered dangerous vermin until people recognized that they lured tourists there and should be counted as money-makers; the number-one tourist attraction there is now the lion. "Seeing the lions was really the main thing we were after," said an American doctor just back from a safari to Serengeti National Park in Tanzania. "We drove around and finally found them. But," he continued somewhat disappointedly, "they were just lying there, crawling with flies."

Whether in the Serengeti or in Ngorongoro Crater, also in Tanzania, or in Nairobi National Park in Kenya, Land-Rovers, VW Microbuses, and family sedans prowl the plains morning and evening looking for the big cats. It is an old joke that the best way to find lions in the Nairobi park on a Sunday afternoon is to look for clusters of cars. "They are always being watched," said Hassan S. Mohamed, chief warden of the park. "They are always surrounded by cars." He said that the lions no longer hunt during daylight hours, when visitors are in the park. "As soon as a lion gets up then he has maybe ten, twenty cars following him." It's tough to sneak up on a wildebeest or zebra with a dozen motors idling right behind and maybe children squealing and certainly cameras clicking and whirring.

While others were watching Elsa cavort on the screen—played by as many trained lions as Lassie has stand-ins—or marveling at a broken-down circus lion sleeping in a commercial park, George Schaller, a young American biologist affiliated with the New York Zoological Society, was conducting the first serious, scientific study of how lions really live in the wild. Schaller, who first won fame with his studies of the mountain gorilla published in 1963 and 1964 (detailed in Chapter 4) lived in the Serengeti for three years, and spent some 2,900 hours of observing hundreds of wild lions. His monograph, *The Serengeti Lion* (1972), is a landmark in the blossoming field of high-intensity animal behavior research. Along with thousands of details of lion life, Schaller's findings explode many of the most cherished beliefs about this most fabled of beasts. He has found, for example, that:

1. Lions prefer stealing their food from other predators to hunting it themselves. In many parts of Africa lions get more than half their food by scavenging the carcasses of animals killed by hyenas, wild dogs, or disease.
2. In times when prey is scarce, adult lions monopolize what food is available, often leaving the cubs to starve. One quarter to one third of all cubs die this way. A similar proportion of cubs die following abandonment by their mothers.
3. Female lions do more than 90 per cent of the hunting and killing of prey. The magnificently maned males rush in to gobble most of what's killed before allowing the females to feed. The cubs are last.
4. In a lion pride most of the important leadership comes from the females who maintain a permanent, closed sisterhood. The brawny males are itinerants who attach themselves to a pride for a few months or years, monopolize the food, sire a few litters, and then drift away to a vagabond life until they find another pride without a male or with one that can be pushed out. The male's chief service to his adopted family is to defend it against other males.

As noted at the beginning of this book, perhaps the two most sacred myths about lions and, to a lesser extent, about every form of animal life is that, unlike depraved man, animals do not kill except in self-defense and for food, and they certainly do not murder their own kind. When fighting breaks out, according to widely held popular beliefs, one animal is supposed to offer a traditional gesture of capitulation before any fatal blows are landed. Bunk. Now that individual groups of free-living lions have been studied long enough, it is clear that violence toward and murder of their own kind are very much a part of lion society. Schaller observed several lion fights that ended fatally. He saw males from one pride attack a male in a neighboring pride, killing him, and leaving the pride defenseless so that males from yet another pride invaded, killing and eating the cubs. The maleless females also turned cannibalistic, killing and eating their own cubs. It has often been observed that when a male takes over another pride, it will kill the existing cubs. There is an explanation for this seemingly senseless behavior. Females that are nursing cubs do not come into heat and are sexually unavailable to the new male. When the cubs are gone, natural physiological processes trigger ovulation in the lion and release the hormones that signal her readiness to mate again. By wiping out his predecessor's offspring, a male hastens the arrival of his own progeny and eliminates any competition for them. It is unlikely that lions do this consciously. Rather, it appears that this is an innate pattern of behavior which has evolved simply because it is the murderous lion's genes that survive while those of its predecessor, which may or may not have been the murdering kind, perish.

Schaller's accounts of violence among lions in the Serengeti park do not make pleasant reading, but perhaps a small dose of what lions can do to one another is needed to balance the popular image of these beasts. Schaller describes an incident in which three males from the pride he called Masai (named for an East African tribe) apparently attacked one of the two males in the neighboring pride which he called Seronera (for a nearby settlement consisting mostly of tourist accommodations). Although he did not witness the battle, Schaller was aware of increasingly tense relations between the prides following an attempt by the three Masai males to steal a gazelle carcass from females of the other pride. One night Schaller left the lone Seronera male resting not far from the three Masai males. Here is how he describes what he saw the next day:

"The following morning at 0635 the Seronera male was lying in the same place, covered with blood. Tatters of his yellow mane were strewn over an area of three by ten meters. His right eye was closed and a deep gash angled across his left brow. His left flank was ripped to the bone and a hole eight centimeters wide penetrated his chest. Other deep punctures and cuts covered his body, particularly his rump. A bite in the top of his head had broken the sagittal crest [part of a skull bone]. He breathed heavily.

"At 0655 female A walked up to him and seemingly sniffed his mane. Suddenly she growled and fled. Standing 40 meters away was Black Mane of the Masai pride [one of the attackers]. He advanced slowly to within one and a half meters and faced the wounded male which, with his chin resting on his forepaws, gave a brief, low growl. Black Mane scraped the ground with his hind paws and walked to the remains of a zebra in the nearby thicket."

An hour and a half later the Seronera lion, having been able to do no more than lift his head briefly, died. Later that day one of the Masai males mated with a Seronera female. Over the next few months the Seronera females associated with the Masai males and shared kills with them. After the death of one of the two Seronera males, the other "seemed to lose his assurance," Schaller observed. The Seronera pride's social order was breaking down. When the remaining male heard males from other prides roaring, he would walk silently in the opposite direction instead of answering in the usual fashion with another roar. Males from three other prides repeatedly penetrated the Seronera pride's territory. One of the Seronera females killed three of her pride's cubs, biting them to death. Once the Seronera and Masai females happened to meet and there was a great deal of snarling, growling, roaring, and clawing. "One snarling lioness," Schaller wrote, "chased a Masai pride female and clawed her deeply in the rump, an injury which.caused the thigh to wither in the following month until she could only hobble on three legs."

Schaller recounts many other incidents he witnessed in these and many other Serengeti prides, enough to make it clear that while murder and violence are not everyday occurrences among lions, they are not uncommon.

Whether neglect of cubs can be considered murder is an open question, but this again happens frequently among lions. Sixty-seven per cent of lion cubs die before they are old enough to fend for themselves and the two leading causes of death are killing by other lions and abandonment.

Another myth about lions is that they kill only as much as they need to eat. Again, bunk. Lions sometimes go on killing binges, wantonly catching and killing almost any animal they can. "Lions sometimes kill more prey than they can consume in a meal," Schaller says. "At . . . times lions already have a kill but take advantage of an easy opportunity to obtain another, no matter how gorged they are. When large herds of wildebeest migrated past Seronera for several days, a group of lions killed repeatedly, leaving uneaten carcasses lying around." During a drought in Nairobi National Park, Mervyn Cowie, first director of Kenya National Parks, observed lions killing large numbers of starving and weakened animals without eating most of them.

For centuries, the Western world's view of lions was a gentler one, derived from a description by the Roman naturalist Pliny the Elder, which echoes through many of the bestiaries and natural history encyclopedias so popular in Europe in medieval and later times. "The lion," Pliny informed his readers, "alone of all wild beasts is gentle to those who humble themselves before him and will not touch any such upon their submission, but spares whatever creature lieth prostrate before him." What better symbol for a political ruler who wishes himself to be seen as powerful but merciful?

Long before Pliny, the lion had captured the hearts and, in some places, the souls of men. Lions were worshiped in Mesopotamia, Phoenicia, and, especially, in ancient Egypt where the cities of Leontopolis and Heliopolis were centers of a lion cult. In the temple of Amun-Râ at Heliopolis, for example, priests tended tame lions, bathing them in perfumed water and soothing them with incense and sacred music.

To the extent the lion was revered, so its ritual death sanctified an occasion. Julius Caesar, for example, consecrated the opening of his forum in Rome with the slaughter of four hundred lions. Always sensitive to the entertainment value of bloody violence, the Romans made effective use of lions in the circus, often releasing them into the arena with bulls and bears and, of course, condemned criminals. As late as A.D. 1459 the Tuscan city of Florence tried to revive the ancient tradition by barricading all the streets of the piazza and, as a huge crowd watched, releasing into it twenty-eight lions and an undisclosed number of bulls. It was to have been a glorious spectacle, but in fact the well-fed lions simply milled about for a while, ignoring the bulls, and eventually lay down in a shady corner and went to sleep.

The nobility or sacredness of lions has never prevented their being hunted. Indeed, sport hunters have long held that their quest for a particular beast was homage to it and that a bullet or arrow conferred a kind of immortality. Many a modern sport hunter enters his trophy room as if it were a sacred pantheon.

For all the zeal and myths that visiting lion hunters have brought to their sport, those who live in lion country—the civil population and game wardens—have generally held a less enthusiastic view. It was not so many years ago that the beasts now considered part of Africa's "priceless wildlife heritage" were thought to be of no value at all. "One of the least known but most interesting stories of how a long-standing bush belief [about wild animals] could be replaced by its complete opposite," wrote Alistair Graham, a wildlife biologist and former Kenya game warden, in *The Gardeners of Eden* (1973), "is the rise to fame of lions and leopards, a story wholly devoid of the romance modern game savers attach to these beasts. In the original 1904 game laws leopards, lions and cheetahs were not even classified as game animals worthy of sportsmen's serious attentions. On the contrary, they were . . . considered a positive nuisance—vermin." As Graham shows in his study of the rise of the conservation movement in East Africa, the early colonial game departments viewed lions simply as killers, responsible for sometimes hundreds of human deaths each year and for the destruction of the valuable game animals that sport hunters preferred, such as antelopes.

Not until the 1920s, when the number of hunters visiting Africa from Europe and America began to grow considerably, were lions added to the list of species for which shooting licenses were required and the rituals of sportsmanlike hunting enforced. Until then, anyone was free to shoot as many lions as he liked, and it was not uncommon for hunters to return with sixty or more. Few people objected. The more that died, the better the hunting for real game such as kudus, impalas, and gazelles.

There was never much disagreement that lions could be dangerous. But apparently lions did not behave dangerously enough to make a good story for the tourist hunter who had paid thousands of dollars (or pounds or francs, etc.) to come to Africa. Consequently, some hunters resorted to special tactics to make the lion charge so that the bullet could drop it in midleap. Teddy Roosevelt, on a hunting trip after leaving the White House in 1909, favored a method called "galloping the lion." The hunter would use horses and once he had found his target, he would walk and gallop the horse, back and forth, at first only teasing the lion but gradually working it into a furious rage. One bullet, or, more typically, a fusillade, later, the hunter had his lion and a story that would prove anyone's manhood.

Of course, there were plenty of shooters who thought it safer to bag the lion some other way. In the Serengeti Plain, before it was a park, lions were so plentiful and so unafraid of visitors, that it was easy to knock off as many as one wanted. "Lion hunting in the Serengeti became a real farce and no self-respecting hunter would dare to admit that his skins came from that region," recalled C. A. W. Guggisberg, a British zoologist and long time student of lions who, until Schaller, was probably the world's leading authority on lions.

Lions are still common in Africa and nowhere near the endangered species list. But hundreds of them are killed each year and lion hunting continues to be big business in Africa. Interestingly, however, the old low opinion of wild animals once held by many Africans is beginning to change in the cities. There are now millions of Africans who have grown up in agricultural regions or in urban and suburban areas who have never seen wild animals at all or only from the safety of a vehicle. Among these Africans a constituency is developing that favors wildlife conservation, and not long ago the Nairobi newspapers carried dozens of articles and letters to the editors criticizing one Leo Roethe, a fifty-eight-year-old American millionaire who came to Kenya to shoot a lion. In the process, he only wounded the beast which, after five hours of more bullets and harassment, turned and chomped on Roethe's foot. An accompanying professional hunter (sport hunters are allowed into the bush only in the company of a licensed professional, for just these reasons) fired a seventh and fatal bullet, apparently having given Roethe as many chances as he reasonably could to claim the kill for himself. As Roethe waited in a Nairobi hospital for his wounds to heal, the newspapers headlined his dramatic ordeal, pointing out that he was president of an American club called the Organization of African First Shooters which espouses hunting as a prime conservation method. What really fired the ire of many Kenyans, however, was the fact that this was Roethe's thirteenth lion and that back home he had a private museum containing 575 stuffed animals "which I have shot myself." That kind of conservation, letter writers and columnists said, Kenya could do without.

Though the letter writers typify a growing body of Africans who favor traditional philosophies of conservation, many are not ready to embrace the kind of lion adulation that has been triggered by Joy Adamson, as a result of her story of Elsa, a lioness. Elsa today is bigger in death than she ever was in life. Hers is the most famous and perhaps the most bizarre story in the annals of the lion cult. In every sense but one, Elsa was the creation of Joy Adamson, the wife of George Adamson, a British game warden in Kenya. Conservation workers in Africa and elsewhere credit her books and the movie patterned on them with awakening a wide interest in protecting dwindling numbers of wild animals around the world and with eliciting substantial contributions for conservation along with more modest ones. Not long ago, for example, Mrs. Adamson received a check for $90 from a group of children who had formed an Elsa Club in Los Angeles. They had collected old newspapers and bottles, sold them, and sent the proceeds to help save the flamingoes in Kenya's somewhat polluted Lake Nakuru.

"Elsa died in 1961 but still she grows and grows," Mrs. Adamson said in an interview in her Kenya home, which is furnished with chairs upholstered in lion skins. Though now in her sixties and arthritic, Mrs. Adamson said she was busier than ever coping with the continuing worldwide response to Elsa. As a result of the books, she is invited to speak at conservation meetings around the world and speaks with pride of having been invited to visit the Soviet Union and discuss conservation programs there.

Although the *Free* books are widely criticized as overly anthropomorphic, Mrs. Adamson contends that she does not attribute her own emotions and thoughts to the lions. Rather, she says, it is quite the reverse, for Mrs. Adamson believes the lions communicate with her through mental telepathy. "Lions have this extra sense and so do some people," she said. "I have known this kind of thought-communication with the animals with whom I've lived," she once wrote. Mrs. Adamson said she believes Elsa speaks to the world through her and her books. "I believe she came here to do a big job," Mrs. Adamson said of the lioness whose mortal remains lie buried under a landscaped cairn of rocks in northern Kenya.

"Something urgent had to be done to save the wildlife," Mrs. Adamson said. "There is something about lions you cannot escape—their incredible personality. People say they act so human. I only wish we could act more like lions. They're so noble," Mrs. Adamson said. "The lion knows he's boss, so he can afford not to be petty-minded. I love all animals but I don't think any compare with the lion. It's just . . ." Mrs. Adamson stopped, unable to translate the emotion that animated her face.

"I'll tell you," she resumed, "human beings are the only species that have prostitutes. Animals are very selective in their partners. All the wild animals are really pure. We have lost our purity. Elsa was clean and pure. I'm told that Clark Gable, when he was dying, wanted Elsa's books. After a life like his, he wanted something clean and beautiful."

What manner of lion was this whose death in 1961 made the obituary columns all over the world and whose passing prompted Nelson Rockefeller, among thousands of others, to write a letter of condolence to Mrs. Adamson? As a tiny cub Elsa became an orphan because George Adamson killed her mother, believing her to be a maneater. The Adamsons shipped Elsa's two littermates to the Rotterdam Zoo where, Mrs. Adamson said, "they live in splendid conditions. I was glad to know that almost certainly they had no recollection of a freer life."

Elsa was kept as a house pet, initially much as anyone

keeps an ordinary cat, but eventually as more of a family member.

Although the Adamsons' repeatedly stated goal was to return Elsa to a truly wild and free life, the main adventures in the books are the Adamsons' continual attempts to retrieve Elsa or her cubs after they wandered off into the bush. Long after Mrs. Adamson wrote, "She had proved that she could fend for herself and was independent of us at least so far as food was concerned," the Adamsons continued to kill antelopes to feed Elsa and lure her back from the wild. Eventually, like a Central Park pigeon, Elsa became conditioned to expect a handout whenever the Adamsons arrived in the bush.

The Adamsons' pet lions became such a nuisance, killing the cattle, goats, and sheep of neighboring herdsmen, that Kenya officials ordered the Adamsons to round them up and take them someplace else. Officials in Tanzania agreed to let them release the cubs (Elsa had since died) in the Serengeti National Park. But the officials were aghast to learn that not only had the cubs moved into the park, but the Adamsons had too and were making regular trips outside the park to shoot animals to feed the cubs. Park officials, knowing that a well-fed lion is not likely to learn to kill for itself, thereupon forebade feeding the cubs. Before long the little lions adapted and wandered off to make their own lives. But, desperately worried that the cubs could not survive without their help, the Adamsons spent nineteen months searching for their pets before giving up.

After helping make the movie version of *Born Free,* George Adamson attempted to rehabilitate, as they say, two of the twenty-four tame lions that had performed in the film. (The lion actors, incidentally, were flown from American zoos and other places to Africa.) In a tract of rented land in northern

Kenya, he cared for the pair, named Boy and Girl, as they shed their tameness and became increasingly wild. According to Alistair Graham, then a Kenya game warden, the effort was proceeding well until Boy killed one of Adamson's employees, dragging the corpse back to camp and eating it there. The experiment was, apparently, too successful; Adamson shot Boy.

Strange as the Adamsons' story may be, it at least makes more sense than that of another lion detamer and author, Norman Carr of Zambia. Because wild lions will often attack tame lions in their territory, Carr begins his rehabilitation process by shooting the wild lions in the area where he wants the tame ones to set up housekeeping.

While books like the *Free* series and Carr's *Return to the Wild* (1962) do reveal something of the nature of lions, they tell us little about how natural, free-living lions behave. That is why Schaller's 1972 monograph is such a landmark after the centuries of myth and hunters' tales. One erroneous notion that it corrects, for example, is that lions kill by leaping onto the animal's back and, with a swat of one mighty paw, breaking the victim's neck. Artists' conceptions of this are rampant in some of the poorer wildlife books. Death of the prey, the belief holds, is swift and clean, even merciful, as befits a noble creature. The lion's method is held to be more admirable than that of hyenas or wild dogs, perhaps the two most despised animals in Africa. Unlike the noble lion, it is said, they kill in mobs by cruelly running down their prey and nipping relentlessly at its flanks until fatigue and loss of blood bring the creature to an end. Typically the prey is killed by disembowelment.

The British naturalist Jane Goodall and her then-husband, Hugo van Lawick, a Dutch photographer, studied wild dogs over a two-year period and concluded that, contrary to the popular notion, wild dogs are "quick and efficient killers," often more so than lions. "Throttling, the system used most often by the cats," they said, "is less gory and is, therefore, considered to be a 'kinder' way of killing; sometimes, though, it takes ten minutes for a victim to die." Goodall and Van Lawick found that wild dog kills nearly always took less than two minutes.

In the hundreds of kills observed by Schaller and other reliable witnesses, only once did a lion leap at its quarry and never did Schaller find any evidence of broken necks. Most typically a lioness kills by quietly sneaking up on its victim until it is only ten to thirty yards away and then, with the advantage of surprise, bursting out of hiding and running at the prey. Most often, Schaller found, the relatively slow and heavy lions bungle the job by charging from too far away or zigging when the prey zags. But should the predator get within paw's reach, it hooks its claws into the animal's flank, throwing it off balance and bringing it down. Then the lion bites the animal's neck, usually at the throat, and keeps its teeth clenched until the animal strangles. Sometimes a lion will actually place its mouth over the muzzle of the prey, suffocating it. Never does the lion growl or snarl at its prey; such aggressive display is reserved for fellow lions. The kill is made as dispassionately and as quietly as a homemaker buying a roast.

When lions hunt as a group, sometimes using complex pincer tactics, Schaller has found that they are twice as likely to make a kill as when hunting alone. This, he theorized, was probably a strong factor favoring evolution of the pride as a social unit. Lions are the only members of the cat family to live in groups, all the others being able to hunt and kill success-

fully as individuals. Seldom is there any heroic struggle between prey and predator. Curiously, once an animal has been
pulled down, it usually stops struggling. As if in shock, it lies
still while the lions rip into the flesh. Schaller cites an extreme
case in which a buffalo, pulled down but not badly injured,
lay quietly on its side while a lioness chewed on its tail.

Once an animal is subdued, Schaller wrote, "a kill has a
most disruptive influence on lion society. The lions seem to
become asocial, as each animal bolts its meat while snarling
and slapping at any group member that seems to threaten its
share. Crouched or standing, lions surround a kill and eat
wherever they can in a writhing and growling mass until the
carcass can be dismembered."

Of all the food lions eat, surprisingly little is killed by the
lions themselves. Often they steal it from other predators such
as hyenas and wild dogs, species whose popular images are as
undeservedly low as the lion's is high. Hans Kruuk, the Dutch
zoologist whose studies on hyenas rival Schaller's on lions,
says, "It is clear that . . . lions more often make use of food
killed by hyenas than the other way around." (A fuller discussion of Kruuk's findings about the hyena appears in Chapter
8.)

Kruuk and Schaller independently calculated the scavenging rates of lions and hyenas in the Serengeti and found that
of the cases in which one scavenged from the other's kill, 54
per cent of the kills were made by hyenas and 34 per cent
were made by lions. In the remaining cases it could not be determined which had killed the prey. Kruuk and Schaller agree
that if a lion can find its food already dead or steal it from another predator, it will not hunt and kill for itself. To help
them find scavengeable food, lions have learned to watch for

vultures circling in the sky. They will often walk a mile or more to find the carcass below the vultures.

Of course, lions would rather not have to walk at all. They would prefer simply to lie down in some comfortable spot and sleep or, at the most, stare into the distance and twitch the flies off now and then. Schaller calculated the average time that a lion spends doing various things and discovered that the primary activity is inactivity. Lions spend between twenty and twenty-one hours a day sleeping or resting. They spend roughly two hours walking (not all at once) and about forty to fifty minutes a day eating. Only in the remaining few minutes is the lion earning its living—stalking prey and making a kill or, likely as not, chasing hyenas off their kills. What this means, of course, is that the way you see lions at the zoo is pretty much the way they are in the wild. Although lions, cubs especially, have a reputation for being playful, and such behavior is not uncommon in zoos, free-living lions are rarely so. Schaller observed three groups of lions for play behavior and found they spent an average of only 1.5 per cent, 1.6 per cent, and 6 per cent of their time in play.

Along with the recent findings of other researchers in Africa, Schaller's work goes a long way toward demythologizing and demystifying the lion and putting him back into his proper ecological niche. As the world-wide desire to preserve the animals of the wild continues to grow and begins to dictate wildlife management priorities and policies, there becomes less and less room for false or sentimental notions about any species. (There was a time when East African national park authorities had to draft a serious reply to a demand that, in keeping with the idea of preserving wildlife, park rangers should prevent lions from killing other animals.)

Hyenas and wild dogs are not the wretched, filthy creatures

most of us think, and neither are lions noble and majestic animals ruling over their African domain. The lion is pretty good at hunting, but it botches the job much of the time. It contributes to the general welfare by weeding out the sick and aged prey animals, but it also kills healthy, defenseless baby antelopes and zebras. Lions prefer to live peaceably and they have turned laziness into a fine art, but once in a while they fight among themselves and viciously kill one another. In other words, the lion, like all other creatures, is simply an ordinary working stiff, trying to make a livnig the best way it can.

3

WOLVES

Strange tales have come to us from the land of the frozen north but few have seemed more believable or more vivid than the stories of Jack London (1876–1916), that master of the adventure narrative. Generations of school teachers have told children that he could write about the north woods so well because he had lived there and experienced the things he wrote about.

"Dark spruce forest frowned on either side of the frozen waterway," London wrote at the beginning of his novel *White Fang* (1906). "The trees had been stripped of their white covering of frost, and they seemed to lean toward each other, black and ominous, in the fading light. It was the Wild, the savage, frozen-hearted Northland Wild."

At the bottom of this wintry realm two men, Bill and Henry, were accompanying a dogsled with a coffin, plowing across the snow day after day to deliver a corpse to its family. Behind the men trailed a pack of hungry wolves that, each night, would steal into camp to attack and eat one of the dogs.

One night the two men were huddled near their campfire: "Bill opened his mouth to speak, but changed his mind. Instead he pointed toward the wall of darkness that pressed about them from every side. There was no suggestion of form in the utter blackness; only could be seen a pair of eyes gleaming like live coals. Henry indicated with his head a second pair and a third. A circle of the gleaming eyes had drawn about their camp."

Night after night the wolves pressed in with the darkness and each night a she-wolf would approach the dogs, enticing one toward a suggested liaison that invariably ended in death as the whole pack leaped snarling and slashing at the would-be suitor. Eventually Bill and Henry were down to three dogs and the wolves were now bold enough to approach during the day. When the ravenous beasts took out after one of the last dogs, Bill grabbed his rifle and their last three rounds of ammunition, determined to prevent another killing. His three shots missed and the wolves set upon him as well as the dog. Within moments there was only silence, but still the pack was hungry. Henry, now alone, began preparing early to survive the night.

"But," as London wrote, "he was not destined to enjoy that bed. Before his eyes had closed the wolves had drawn too near for safety. It no longer required an effort of the vision to see them. They were all about him and the fire, in a narrow circle, and he could see them plainly in the firelight, lying down, sitting up, crawling forward on their bellies, or slinking back and forth. They even slept. Here and there he could see one curled up in the snow like a dog, taking the sleep that was now denied himself. He kept the fire brightly blazing, for he knew that it alone intervened between the flesh of his body and their hungry fangs."

London's story goes on, with the allegedly bloodthirsty beasts actually trying to attack, "the jaws snapping together a scant six inches from his thigh." Eventually, after driving off the marauders by hurling burning brands, Henry is rescued in the nick of time.

Jack London could tell a gripping story all right, but he really didn't know much about wolves. Rather than relying on first-hand experience for his writings, London borrowed instead from fairy tales. There is, for example, no evidence of a wild wolf ever killing a human being in all of North America. There is only one documented case of a wolf even injuring a person and that animal almost certainly had rabies. Even in Europe and Asia, where the wolf has been the archetype of evil for thousands of years, modern authorities on wolf behavior have concluded that the vast majority, if not all of alleged wolf attacks are false and that the handful of proven attacks were the work of rabid wolves or of wolf-dog hybrids.

In fact, as a rapidly growing body of scientific information on wolf behavior is showing, the wolf is actually a rather shy creature with a strong aversion to fighting. It has a rather playful, friendly nature among its fellows. Research findings to date show wolves to exhibit many of the behavioral patterns that should find favor among the more sentimentally inclined animal lovers.

The wolf pack is a tightly knit family in which co-operation rather than competition is the rule. Wolves are frequently monogamous; both males and females take great interest in raising and playing with the young; unlike lions, which may not leave their cubs much food in time of famine, adult wolves readily disgorge predigested meat for pups whenever the little ones indicate their hunger even in times of famine; when a mother wolf goes out on a hunt, another will babysit with her pups. The wolf howl, long portrayed as a deliberately terrifying signal of impending death, is now known to be an important communications medium among wolves, helping separated pack members to find each other and serving as recreation during group howls where even the little pups join in, tottering around with wagging tails and giving out high-pitched yips.

This is not the wolf Jack London knew or apparently even heard about in his brief exposure to wolf country. Despite his reputation, Jack London spent no more than a single winter in the Canadian north woods—futilely searching for wealth in the Klondike gold rush of 1897–98. No seasoned woodsman, London was a greenhorn who once eagerly rushed into Dawson City with a bagful of glittering mineral, intending to register his claim. At the assayer's office London learned he had struck mica. On another occasion London's clumsiness with an ax destroyed the cutting edge of a particularly fine one he had borrowed and nearly destroyed his friendship with the man who owned it.

Even though London had so little opportunity to learn what wolves are really like and even though his literary motto was "Make it vivid. Truth doesn't matter so much, so long as it lives," London can hardly be accused of deliberately falsifying his wolf stories. Rather, it seems likely that Jack London first learned about wolves like most of the rest of us—from the tale of Little Red Ridinghood.

In this beloved children's story the wicked wolf, as you may recall, first eats up Grandma and then disguises itself in the old lady's nightgown to deceive Little Red Ridinghood. Young Miss Ridinghood is fooled and also devoured by the insatiable monster. Then comes along a brave hunter who slays the wolf and slits him open to let the two victims step out.

The wicked wolf also figures prominently in such other children's classics as Walt Disney's *The Three Little Pigs* and Serge Prokofiev's *Peter and the Wolf*.

The wolf's bad reputation goes back farther than that, of course. It goes back at least to the days of Aesop who had no fewer than six fables about wolves, the most famous being "The Wolf in Sheep's Clothing." This is about a wolf that, in order to get close enough to a flock of sheep to make a kill, dons a sheepskin disguise. A few hundred years after Aesop, the story was picked up by one of the biblical authors who wrote (Matthew 7:15), "Beware of false prophets which come to you in sheep's clothing, but inwardly they are ravening wolves."

It is probably true that throughout history no animal has been more hated or feared, certainly none has been more widely and deeply misunderstood, than the wolf. To a great extent this is because no major predator has lived so close to so many human beings in so many parts of the world.[1] Before

[1] Hyenas, which once flourished in Europe, died out there long before Europe was thickly populated. The only exception to this would be bears. If one lumps together seven genera of bears, they can be said to live wherever wolves do and in South America as well. For several reasons, however, bears (except grizzlies) have a positive image—cuddly, cute, etc.

man's settlements grew too large, the wolf occupied virtually the entire area of North America, Europe, and Asia, including India—then the three most populous continents. Although there were varying colors (from black to white) and sizes (40 pounds to 100 or, rarely, 150 pounds) of wolves throughout this vast territory, all the animals called "wolf" have been members of a single species. Biologically this is unusual among mammals because differing habitats usually present such different challenges to once similar animals that they gradually evolve distinctive adaptations to their locales. The two most obvious exceptions to this pattern are wolves and human beings—both for more or less the same reasons: adaptability to a variety of habitats and a habit of roaming and migrating to meet and mix with neighboring groups.

The contact between man and wolf almost certainly stretches back to man's early days as a hunter in Europe and Asia. The evidence is the dog, all varieties of *Canis familiaris* being the descendants, most authorities agree, of domesticated wolves. At some time, perhaps many times, in man's past some of those pairs of "glowing coals" that pressed around the campfires at night overcame their shyness and padded into the company of men and women and children. There would have been little incentive to reject the intruder. Before we had domesticated cattle or goats, wolves could not have engendered man's hate by stealing them. When we were hunters, the wolf's way of life may well have seemed a model. And no one had yet invented the myth of the man-eating wolf.

Whether the first pet wolf was an adult or, more likely, a tiny pup, wolf and man became inseparable. Countless centuries of selective breeding and a taste for oddly formed mutant offspring have led to the more than four hundred different varieties of domestic dog known today. All retain enough of the wolf in them to permit them still to mate with wolves and produce crossbreeds. Nearly all also display a variety of wolfish behavioral traits, the two most notable of which are the wolf's overwhelming desire for social contact—making friends and becoming loyal, devoted companions—and an aversion to fighting among their friends. Strange as it may seem, the qualities that have made dogs such popular pets are the very behavioral traits that ally them most closely with wolves. This fact had remained forgotten through thousands of years of hatred and persecution of the wolf until a handful of biologists in the last thirty years (and chiefly in the last ten years) discarded the myths and relied on objective methods of observation of wolves to learn what these creatures are really like. Through a small but growing number of books and articles based on their studies, these biologists have slowly fostered a new appreciation for the real wolf.

The father of modern wolf research is the biologist Adolph Murie, who published his study of the wolves of Mount McKinley in Alaska in 1944 when he worked for the United States National Park Service. "The strongest impression remaining with me after watching the wolves on numerous occasions was their friendliness," Murie wrote in a monograph published by the Park Service. "The adults were friendly toward each other and were amiable toward the pups. This innate good feeling has been strongly marked in the three captive wolves which I have known."

The leading modern researcher on wolves is L. David Mech, whose 1970 book, *The Wolf*, is perhaps the best single volume on the animal. "Probably the creature's strongest personality trait is its capacity for making emotional attachments to other individuals," wrote Mech, who is now a biologist with

the United States Department of the Interior's Fish and Wildlife Service.

Through the work of scientists like Murie and Mech, who
have elucidated the details of wolf life, it is now possible to
understand wolf behavior to a much greater extent than before through the patchwork of sporadic and accidental observations by untrained witnesses.

Wolves live in highly developed social groups called
"packs," which include anywhere from two or three animals
to ten or fifteen in a large pack. Very rarely packs comprise as
many as twenty individuals and there is a single reliable observation of a pack of thirty-six. The storyteller's "howling
mob" of hundreds of wolves in a single pack is pure fiction.
Ordinarily a pack consists of a breeding pair of adults and
one or two litters of their pups, the oldest of which are often
young adults. As the youngsters mature, around the age of
two, they usually go off to form packs of their own, keeping
the size of the typical wolf pack at seven or fewer animals.
Within the pack there is a strict hierarchy, or pecking order,
in which each wolf knows who is dominant over it and which
lower ranked animals it may dominate. Because a wolf's survival depends heavily on the ability of its pack to hunt and
kill co-operatively, the daily life of a wolf is filled with minor
acts and ceremonies that maintain friendship and solidarity
within the pack. For example, whenever any two members of
a pack are separated, even if only for a matter of hours, the
subordinate animal eagerly rushes up to nuzzle, lick, and nip
the mouth of the dominant animal. The ceremony appears to
reaffirm the subordinate's acceptance of the pack's all important social structure. When your pet dog does the same thing,
leaping up to slurp at your face, it is merely acknowledging
you as its superior, reaffirming its friendship and loyalty. Just

as the wolf's survival depends on teamwork with its superiors, your dog relies on you to keep his stomach full. If your dog plays with other dogs in the neighborhood, he may well have formed similar dominant-subordinate relationships with them. Careful observation may reveal whether your dog is a leader or a follower in his crowd.

The second wolf trait shared by dogs, if not quite so widely, is an aversion to fighting. The two premises on which wolf society is based are organization and co-operation. Without it, the chances of killing a moose or an elk several times the size of a wolf are much smaller. Survival demands co-operation. Dogs who depend only on canned dog food have less need to avoid fighting except with their owners, from whom they sometimes take brutal punishment without retaliating. The extent to which a dog is a muted wolf can be seen in abandoned dogs. They quickly form packs with dominance orders quite like those of wolves.

Wolves are so averse to fighting anything except a prey animal that they tolerate extraordinary interference in their lives. Murie, for example, once described an incident when he approached a wolf den where a mother and pups were hiding. When Murie was about twelve feet away, the mother darted out and Murie crawled in to snatch a pup. While he maneuvered around the dark tunnel and emerged carrying a tiny, wriggling pup, both mother and father could bring themselves to do nothing more than whimper and bark from the sidelines.

This episode, incidentally, is evidence contradicting the popular belief that animal mothers protect their young with every resource at their command. Another incident, involving wolves in northern Canada, also puts the lie to that notion.

David F. Parmelee, a zoologist at Kansas State Teachers College, was camping in the wilderness of Ellesmere Island, in the Arctic Ocean west of Greenland, with a friend when they came upon a wolf with four pups. The two men chased the wolves for a mile and caught two of the pups. On their way back to camp with the pups, Parmelee shot some ptarmigans for museum specimens. As darkness fell, the two men were walking along, each with a baby wolf cradled in an arm and several ptarmigans slung over a shoulder. They heard something. "Following close in my footsteps," Parmelee wrote, "was the big she-wolf, her nose touching the ptarmigans as they swayed back and forth. Incredible as it surely is, we several times had to drive that wolf off with snowballs for fear that we would lose our specimens!"

Many wolf researchers have reported a variety of similar situations in which wolves offered no resistance whatever to human beings meddling in even the most disruptive ways with wolf life. Mech has live-trapped many wolves for examination and tagging, finding that all he had to do to keep the beast steady was to grab the animal by the scruff of the neck. They seldom offered any resistance.

If wolves are not man-eaters, or even man-killers, how and why did the idea grow up that they were? The belief undoubtedly originated in Eurasia long ago and there is some reason to believe that it started as the result of the behavior of wolves infected with rabies. This is difficult to prove, of course, for a positive diagnosis requires laboratory examination of the suspected animal's tissues. However, enough is known of the bizarre behavior patterns of modern rabid wolves and dogs to see parallels in the descriptions that have been passed down through centuries of folklore.

For one thing, the wolf that actually has attacked people is generally portrayed as a slavering beast that approaches many people—school children walking home, whole families out rid-ing in sleighs, townsfolk in the streets—and attacks repeatedly, slashing right and left, killing again and again but seldom, if ever, eating its victims. Such a wolf would indeed be terrifying. That is how rabid wolves and dogs behave. Healthy, wild wolves, on the other hand, are shy of human beings and reluctant to attack even when harassed by them. Whenever a healthy wolf kills, it is for food and the carcass usually is eaten almost completely.

Also suggesting rabies in aggressive wolf behavior is the fact that in many of the well-documented cases of wolf attacks nearly half the victims, most of whom were not severely wounded, died anyway. One exception to the nearly 50 per cent mortality rate was a group of nineteen persons attacked in France by a single rabid wolf, all of whom were subsequently treated by Louis Pasteur with his new rabies vaccine. Of the nineteen only three died and one of them had suffered substantial blood loss before treatment. Quite probably the rabies vaccine—an early version of the one used today—saved the victims' lives.

C. H. D. Clarke, a Canadian biologist interested in wolves and chief of the Fish and Wildlife Branch of Ontario's Department of Lands and Forests, has reviewed many of the cases of wolf attacks from Europe and concluded that "The famous wolves of medieval song and story were all rabid." He believes the same holds true for wolves attacking in more recent times with one major exception, the so-called Beast of Gévaudan, a creature thought to be a wolf that terrorized the French province of Languedoc by killing nearly a hundred persons between July 1764 and June 1767. The beast's reputation was given wider currency thanks to Robert Louis Stevenson's *Travels with a Donkey in the Cévennes* (1879). In this work, Stevenson tells of a wolf that had been terrorizing the countryside in south-central France by killing person after person. One day the beast attacked a beautiful shepherdess, and, according to other accounts, as hunters tracking the animal approached the scene of the crime, they found a trail of clothes evidently ripped from the maiden's body. Finally they came upon the poor, dead victim, naked and bloody from wolf bites on her neck and thigh. (Some wolf experts, but not all etymologists, credit this tale with being the source of the term "wolf" to describe a man whose intentions are to attack and ravish young women.)

When hunters in Languedoc finally slew two wolves thought to have been the marauders and the killings ceased, the dead animals were taken to Paris for examination and exhibition. There is therefore considerable description in various historical records of their appearance. Both animals had unwolflike coloration and markings that, Clarke concluded, suggest they were not purebred wolves but crossbreeds with domestic dogs.

Quite probably the wolf's reputation grew out of lurid stories of the attacks by wolves in a day when the true nature of rabies was unknown. Rabies will, of course, turn any child's gentle Rover into a vicious killer.

In Russia, where the wolf has long borne an evil reputation, several authorities now admit that the tales have been exaggerated. In 1931 S. I. Ognev, a Russian zoologist, reported that "Cases of attacks are much rarer than is believed. Dinnik [another scientist] reports that he knows of no authentic case of a (nonrabid) wolf attacking a man in the Caucasus." In 1956 G. A. Novikov, another Russian zoologist, wrote that "rabid wolves attack even men, although wolves normally avoid humans and there are fewer authentic cases of attack on man than is supposed." More recently, there have been further attempts in the Soviet Union to document even a small number of wolf attacks. Robert L. Rausch, of the Alaska Department of Fish and Game, discussed the effort with a leading Soviet mammalogist and later reported to David Mech that the Russian "told me that an effort had been made by one of his associates to document some of the reported attacks by wolves. It was impossible to do so, and it was concluded that, with the exception of possible killing of small children wandering alone in remote areas, reports of such attacks had no basis in fact."

More or less the same conclusion was reached in 1940 by James W. Curran, editor of the Sault Sainte Marie (Ontario) *Star,* who for years had offered a standing reward for anyone who could document an attack by a wolf on a human being. The offer was widely publicized throughout North America and even reached a few European countries. None of the claims that came in could offer supporting evidence and Curran concluded, "Any man who says he's been et by a wolf is a liar."

Thus it appears that centuries of folklore to the contrary, there is little, perhaps no, reliable evidence that healthy, purebred, free-living wolves attack human beings.

Still, traditional animal lovers may ask, won't any animal that is wounded or cornered fight for its life? Not necessarily, if it is a wolf.

"That a cornered or wounded wolf is not necessarily aggressive," Mech wrote in his book, "can be vouched for by conservation officer-pilot Robert Hodge of Minnesota who has shot hundreds of wolves from an airplane [when elimination of wolves was an official goal]. Once when he had only broken an animal's front legs, he approached the wolf [on foot] to finish it off. Instead of finding a snarling and threatening mankiller, he encountered a meek and docile creature which wagged its tail in the friendly, submissive gesture of a whipped dog."

This is the animal that has had a price on its head for at least 2,700 years—the earliest recorded bounty having been offered in ancient Greece. For at least 135 generations of man, bounty hunters have been chasing, trapping, poisoning, stoning, shooting, spearing, and clubbing wolves all over Europe, much of Asia, and, for the last three centuries, North America.

When the first English colonists landed on the North American coast, they brought with them an already deeply ingrained attitude toward the wolf. The beast, the myth held, was a vicious killer, dangerous to man and animal alike, and America would not be a good place to settle and raise domestic animals until the danger of wolves could be removed. Records of the early colonies contain repeated accounts of efforts made to guard against or to exterminate wolves. By 1717 community leaders in Cape Cod, for example, had worked up the most ambitious plan yet to make Barnstable county, which was coextensive with Cape Cod, safe from wolves. They proposed to erect a six-foot-high board fence, completely walling off Barnstable from adjacent Plymouth county. The plan failed when the good people of Plymouth objected to keeping all the wolves on their side of the fence.

In 1739 Israel Putnam, later a general and hero of the Revolutionary War, tangled with some wolves on his Connecticut farm. The episode planted the seed of a legend that, particularly in the East, was taught to generations of school children. It is the story of "Old Put and the Wolf." The incident began one night when Putnam was said to have lost seventy-five sheep and goats to a wolf who was too clever and crafty to be caught in conventional traps.

Putnam organized a hunt for the beast that led to a dark, narrow cave where the animal was resting. Putnam's men threw burning straw into the cave to smoke out the wolf so it could be shot. That failed, and Putnam tried burning sulfur but again without success. Finally Putnam descended into the cave himself, to confront the beast personally. As Putnam's biographer described the cave, "It was as silent as the house of death. None but the monsters of the desert [an archaic term for "wilderness"] had ever explored this solitary mansion of horror." Putnam, of course, found the animal and slew it. No doubt the legend of Putnam's bravery would have been diminished had people known then how readily a wolf will submit in the presence of a human being.

Whether Putnam's wolf actually killed seventy-five farm animals in a single night is open to doubt, but there is no question that wolves have been responsible for vast destruction of domestic animals such as cattle, sheep, horses, pigs, and goats. Before the Europeans arrived in North America wolves preyed on deer, bison, moose, and other large wild animals—all of which evolved antiwolf defense mechanisms. As a result, the prey species never lost enough individuals to wolves to reduce its population. In fact, the prey species benefited from the weeding out of sick and weak animals. When the white man killed the wild herbivores or pushed them off the land, substituting his own tame, domesticated, defenseless animals, wolves had little choice but to turn to these for food.

During their presidencies both George Washington and Thomas Jefferson devoted considerable attention to the wolf problem as it affected sheep farming, a budding enterprise that the new nation needed to become economically independent of

England. Washington carried on a long correspondence with European experts on the best methods of eradicating wolves.

There are, in today's wave of wolf enthusiasts, those who will tell you that the historic antipathy of North American ranchers to wolves is based on a misunderstanding of the wolf and an exaggeration of its predation on livestock. The wolf has certainly been misunderstood, but there is no questioning the fact that wolves have destroyed many millions of dollars' worth of domestic animals. "Anywhere [in the remaining territory of the wolf] that livestock is raised," says David Mech, "wolves can be expected to prey on the domestic animals. And it is unreasonable to expect farmers and ranchers to tolerate such predation."

It is unfortunate that the wolf was eradicated from most of the United States but there is little question that it had to be done. A society dependent on agriculture cannot coexist closely with wolves. In recent years the ranchers' concept of "predator control," which means killing predators such as wolves, has been commonly cited by conservationists as one of the stupidities of man's attitudes toward wild animals. While the goal of control may have been reached with inefficient and, at times, counterproductive methods, it is difficult to argue that the small-scale ranchers and farmers of two hundred or fifty years ago should have tolerated repeated substantial losses of their animals, especially the young, to wild predators.

One of the commonest ways of "controlling" wolf populations has been poisoning, usually with strychnine. Ranchers frequently shot bison to use as bait and merely salted the carcass with poison. One of the unwritten "range laws" of the Old West was that no stockman would ride by a carcass of any dead animal without poisoning it. Every good citizen of the range carried a supply of strychnine with him at all times.

The bounty system of wolf control also spread with the westward push of pioneer Americans. Bounties, or cash rewards for bringing in a dead wolf or portion of its body, were paid by local governments and stockmen's associations. As the war on wolves gradually escalated, bounties grew larger, reaching $150 per wolf in some places by 1910. Considering the relative value of the dollar back then, it is hardly remarkable that professional bounty hunters entered the wolf war in increasing numbers. During the winter of 1912–13, for example, one such hunter, a William H. Caywood, brought in 140 wolves and collected $7,000 (the bounty in his area was $50) from the Piceance Creek Stock Grower's Association in Colorado. By 1914 the bounty system was beginning to get out of hand. Over $1 million a year was being paid out and investigations revealed that hunters were sometimes collecting on "scalps" of foxes, dogs, and other animals. What's more, there seemed to be little evidence that the war was having much of an effect on wolf predation upon livestock. Indeed, there was evidence that hunters and trappers sometimes aggravated the problem by merely wounding a wolf and forcing it into stock raiding because it was no longer able to kill the more elusive wild game. Some wounded wolves, such as Minnesota Three Toes, Custer Wolf, and Colorado Stumpfoot, became famous outlaws with bonus bounties on their heads.

As an alternative to bounty hunting, the federal government, which had entered the picture in 1907 by publishing *Directions for the Destruction of Wolves and Coyotes* (which offered the following: "Hint: find the dens in spring and destroy the pups just after they are born"), reactivated the old Cape Cod wolf fence idea. Huge areas of national forest were fenced off, the wolves inside the enclosure killed, and the land offered to ranchers for use as wolf-free grazing areas. The plan was only moderately successful.

By 1915 a bureaucratic reshuffle handed wolf control over to the U. S. Biological Survey of the Fish and Wildlife Service. This agency succeeded in getting from Congress the first appropriation specifically for killing predators. Throughout the western states a vast bureaucracy was established to push for total victory over the wolf. Extermination was the goal. In subsequent years the Biological Survey added to its list of vermin to be gotten rid of the mountain lion, coyote, bobcat, and bear. In 1918 Congress considered the financing of the wolf war so important that it debated enacting a national dog tax "for raising money with which to pay for the killing of predatory wild animals." The idea was eventually dropped and the nation was spared the irony of taxing the descendants to wipe out the ancestors.

By 1929 Congress was complaining about inefficiency in the predator-control program and appropriated an unprecedented $1 million for a more aggressive program that also added to the list of public enemies such creatures as prairie dogs, jack rabbits, gophers, and ground squirrels. President Herbert Hoover signed the bill in 1931.

In 1941 the Fish and Wildlife Service paused to look back on a quarter century of wolf killing—between July 1, 1915, and June 30, 1941—and take credit for killing 24,132 wolves. Stanley P. Young, then senior biologist with the Service, wrote in a 1942 report that "the wolf has been definitely brought under control."

Indeed it has. The once-unrivaled range of this wild animal in the United States has now been reduced to a small fraction of its former size, and where contact with wolves and wolf predation was once a relatively common occurrence, hardly anyone ever sees or even hears a wolf. Wolf control in Europe and Asia was never as bureaucratized as in the United States, but human expansion was just as effective in eliminating the beasts. The last wolf in England died during the sixteenth century and in Scotland in the eighteenth. In most of the heavily populated areas of Europe the wolf was gone by the early twentieth century, although occasional individuals are seen in mountains and forests. In the United States, where wolves once inhabited the entire country with the exception of coastal California and the general area of Florida, Alabama, Mississippi, Louisiana, and Arkansas, they were virtually exterminated east of the Mississippi River one hundred years ago. About a thousand remain in northern Minnesota. Populations west of the Mississippi were gone by the 1930s, although undocumented reports still allege that remnant groups remain in a few spots.

Yet despite the fears of some wildlife conservationists today, the situation does not appear to warrant despair for the survival of the species. The wolf cannot in any way be considered in danger of extinction. It is true that within some political boundaries wolves have vanished. Within the conterminous United States probably not more than 1,000 remain, the vast majority of these being in northern Minnesota. But there may be as many as 25,000 in Alaska and another 17,000 to 28,000 in Canada living all the way into the Arctic.

"In several European countries," Mech reported in *Natural History* magazine, "where much of the populace resides in small rural villages and no longer persecutes wolves so intensively, wolves live in proximity to people. They may wander through a village at night and snatch up the local dogs or even visit the outskirts of a large city such as Rome. In North America they are quickly shot or trapped if they do this."

In the western part of Europe modest but apparently stable populations of wolves live in the mountains of Spain, Italy, Sicily, Poland, and Czechoslovakia.

Still larger numbers of wolves breed throughout Eastern Europe, in the mountains, forests, and plains of Yugoslavia, Romania, Bulgaria, Albania, and Greece. And in the Soviet Union and China wolves are common throughout most of the land.

All the world's wolves belong to the same species, divided into many subspecies, or races—depending on superficial and sometimes arbitrary differences such as fur color and size—and in complete fairness to purists among the conservationists, it should be said that several of these races are already extinct or deeply threatened. Before widespread persecution of wolves in the United States, there were, in the view of some authorities, seven races of wolf in the conterminous forty-eight states. Today six are gone, for all practical purposes. The one wolf race that remains sizable in the area is *Canis lupus lycaon,* sometimes called the eastern timber wolf.[2] This type, the one in Minnesota, is abundant also in Ontario. Several other races of wolf inhabit Canada and Alaska. Because the differences among races are so slight, there is considerable disagreement among zoologists as to whether the

[2] Wolves (*Canis lupus*) are sometimes called by the type of habitat in which they are found. Timber wolves live in forests, tundra wolves live in the Arctic, and prairie wolves, now extinct in North America, lived in grasslands. The "brush wolf" is a term usually referring to coyotes, a smaller animal than a wolf. In some reference works coyotes are erroneously called prairie wolves. In eastern Texas there is an almost extinct animal known as the "red wolf" that appears intermediate in physical characteristics between the wolf and the coyote and which many experts believe is a hybrid of the two.

concept of wolf races has any scientific meaning—a situation quite parallel with that attending the division of human beings into races. Just as some anthropologists count three races of man and others four, five, six, and seven, zoologists classify wolves into greater or fewer numbers of races. To lament the demise of a single race of wolves may mean no more than to lament the passing of a wolf population in which 73 per cent of the animals were dark gray in color while in surviving wolf races the proportion of gray animals may be 51 or 88 per cent.

Despite many centuries of persecution by man, the wolf has survived, at least thus far, for the same reason it was able to populate virtually all of the Northern Hemisphere in the first place: it is an intelligent, adaptable animal, with a well-organized social structure. David Mech's study of wolves makes this important facet of wolf life vivid.

"One day," Mech wrote, "I watched a long line of wolves heading along the frozen shoreline of Isle Royale in Lake Superior. Suddenly they stopped and faced upwind toward a large moose. After a few seconds, the wolves assembled closely, wagged their tails, and touched noses. Then they started upwind, single file, toward the moose."

Mech and other scientists have seen similar apparently organized behavior frequently. Through years of observation in the wild, analysis of behavioral patterns, and experiments with captive wolves, Mech knew exactly what the wolves were doing in the incident just quoted. Stated in anthropomorphic terms, the wolves were out on a hunt, hoping to see or smell an animal they might kill and eat. They scented the moose and consulted each other about the prospect of going after it. The wolves huddled and agreed that this looked like a pretty good opportunity. They broke out of the huddle and lined up behind their leader to veer from their original path and close in on the moose. Such co-ordinated behavior is, of course, part of a highly evolved social organization that maximizes the chances for survival of the pack.

Scientific studies have found that wolf packs have established leaders that make the final decision on when to hunt, when to rest, when to travel, when to attack, and so on.

The leader is usually the top ranked male in the pack's "pecking order." Ordinarily he is the father of the younger pack members. There is a separate dominance hierarchy among the females with the male leader's mate being the top ranking female. She may also dominate over lower ranking males. Although she does not ordinarily compete with her mate for over-all leadership, she may assume the pack leader's duties for brief periods.

Adolph Murie found it easy to spot the leader in, for example, a pack of seven adults and five pups he studied in Alaska. The leader was the largest male. "The other wolves approached this one with some diffidence, usually cowering before him," Murie wrote. "He deigned to wag his tail only after the others had done so. He was also the dandy in appearance. When trotting off for a hunt his tail waved jauntily and there was a spring and sprightly spirit in his step." Mech, in his study of a pack of fifteen to sixteen wolves on Isle Royale, in Lake Superior, found that the leader usually headed the single file when the pack was on the move. When the leader started running, the others did too. Sometimes when the pack was chasing a prey animal, the leader would whirl around and lunge at the followers to abort the chase. Pack leaders apparently decide when a chase is futile and not worth pursuing.

"When the Isle Royale wolves traveled overland, through deep snow," Mech wrote, "one wolf would break the trail and remain many yards ahead of the others, as though of superior strength and stamina. The whole pack rested when the first wolf did, and began to move again when this animal started. After the wolves slept for long periods, leadership became evident in their waking activity. Usually one animal would arise, walk over to each of the others, and arouse them."

Mech's observations have led him to conclude that the leader of the pack operates within a form of government somewhere between absolute autocracy and open democracy. Sometimes it appears to be one and sometimes the other. When the leader aborts a chase, it appears autocratic, Mech concluded. But he found that at other times the leader seemed to respond to public opinion. "The island's pack of 16 wolves was headed across the ice of Lake Superior from Isle Royale toward the mainland, some twenty miles away," Mech recalled. "The wolves had gotten about one and one half miles from the northeast tip of the island when dissension among them became evident. Although most of the wolves appeared unwilling to continue, the leader seemed determined. Several times he returned to the hesitant pack and apparently tried to urge the members on. They continued for another half mile or so until they came to a section of rough and jagged ice. After testing this, the pack returned to Isle Royale. I had the distinct impression that it was the hesitancy of most of the pack members that stimulated the leader to turn back."

A wolf becomes the pack leader by having prevailed in contests between itself and challengers. The contest may be an actual battle, with biting and scuffling in the dirt or it may be simply a staring-and-growling match. The struggle for status begins almost from the day the pups open their eyes with tussling and fighting though rarely with any serious consequences. Usually within a matter of days to weeks, there is a dominance order in the new litter, and peace generally holds from then on as long as each animal knows and accepts its place. As the young wolves mature, they take their place at the bottom of the ranks of adults but from time to time one may challenge an elder. Generally speaking, wolf society has a classic, linear dominance order, in which the leader bows to no one, the number-two wolf is subservient only to the leader and lords it over all below himself, number three is dominated by numbers one and two but is dominant over numbers four, five, and so on.

Once a dominance order is established, there is ordinarily little dissension in the wolf ranks. All the animals, most authorities have found, devote much effort to maintaining the hierarchy—the foundation of the pack's stability and the key to its survival. When two wolves meet, the subservient animal acts out certain gestures of submission—nuzzling, licking, and nipping the muzzle of the dominant individual or, if the confrontation is a bit more precarious, flopping down on the ground and rolling onto its side.

However, wolf life is not always idyllic. Because the non-combative rituals of submission and domination are so successful in preventing fights, especially with the leader, as Mech put it, "high ranking wolves have little need for their aggressive energy, which then builds up. Apparently to release this energy, the dominant wolves often pick on the lowest member of the pack by pouncing on it and attacking it."

Another foible of the pack leader is that it does not always pay close attention to what it is doing. Once Mech was watching a pack traveling single file, led by its dominant male and female. "The male was primarily interested in mating with the female, so he remained half a body length behind her, which put the female at the head of the line. However, the female appeared inexperienced at leading, for she hesitated and wandered a great deal, and twice she backtracked." The other wolves noticed the leaders weren't leading very well. As the leading couple zigzagged around, the other wolves took short cuts back to the proper trail and then sat, patiently waiting for the leaders to catch up with the followers. As the dominant pair approached, the followers dutifully pulled their tails between their legs in a gesture of submission while the errant couple marched past onto the correct path, tails flying high in the dominant position.

Though life may be relatively peaceful within any one pack, a wolf's friendliness usually stops short of the neighboring pack. Because a pack may roam over an area of anywhere from fifty square miles to five thousand square miles that it considers its own territory and seldom encounters other wolves, there have been few reliable observations of two-pack encounters. In virtually all such encounters, however, the rival wolves showed hostility. Murie once saw an alien wolf approach the pack of five that Murie was studying in Alaska. The resident pack trotted out to meet the intruder, surrounded it, and started biting it. The outsider immediately rolled over on its back to indicate utter submission but the pack continued its attack. The wolf leaped up to run away but was chased and knocked down twice in two hundred yards. The leader chased the intruder still farther and bit it again. When the wolf finally escaped, its hips and tail were bloody. The notion of wildlife idealists that wild animals have learned to prevent bloodshed among their own kind by inventing rituals of submission is true in the case of the wolf only within its own pack.

There is, however, another mechanism wolves apparently use to avoid conflict between packs. If the interpretations of biologists are correct, the smell of urine on a "scent post" can tell a wolf whether other wolves have been in the neighborhood recently. Because the olfactory sense is so well developed in wolves (and dogs)—it has been estimated to be one hundred times keener than man's—wolves can probably recognize the odor of their own urine and that of their pack mates. An unfamiliar urine scent would be a sure tip-off.

A scent post may be any object—a rock, stump, log, or ice chunk—that is regularly used as a urination target by almost every wolf passing by. "Sometimes," says Mech, "several wolves in a pack will wait in line and urinate on the same object, and they may repeat this procedure each time they pass by."

This behavior has been more intensively studied in dogs than in wolves. Probably it has similar meaning in both species. Ordinary dogs that are allowed to roam outdoors have been studied and many make it a regular feature of their daily travels to check every scent post in the neighborhood. In areas devoid of convenient trees or rocks, the fire hydrant often makes a handy scent post. If there are many dogs in the area, every tree and fire hydrant and even the wheels of parked cars may be marked with urine and, therefore, a "must" stopping point for a dog on his daily rounds. Many dogs find it necessary to add their own scents to each post along the way, much to the annoyance of their leashed owners.

John L. Fuller and E. M. DuBuis, Canadian biologists, made a detailed study of the behavior of domestic dogs and reported: "Ordinarily male dogs who are allowed to roam in a village check the scent posts in their neighborhoods at least once a day. The route covered usually encompasses several miles, and two or three hours may be spent carefully 'reading' each stop along the way. Small groups of dogs meet at various points on the route and continue together. Scent post checking may be interrupted while individuals pursue other interests, but the average adult male does not return home until the usual course has been completed."

If it is assumed that wolves can smell as well as their descendants the dogs, the scent post marking may also be useful to them in determining who is in the area.

By far the leader's most important role is to help the pack find food. It is in charge of the hunt. Next to the myth about wolves attacking people, perhaps the most widely held misconceptions about wolves involve the way they hunt. Once they have selected their prey, the story goes, they chase it relentlessly, sometimes running in relays to wear down the terrified victim until they can close in, slashing and snarling, ripping it to shreds. It is also believed that the wolves snap at the fleeing prey's ankles, severing the Achilles' tendon and hamstringing the animal.

The facts are that wolves can't catch the vast majority of animals they chase and are afraid to attack the majority of prey animals they do catch up to. Mech watched 120 instances in which wolves detected moose and began to close in. In these encounters, 113 moose—94 per cent—escaped unharmed, most after only a minute or two of harassment. One moose was bitten but escaped, and only six were actually killed and eaten. Throughout Mech's studies and those of the other wolf researchers there is no evidence of hamstringing. In addition, most of the moose Mech watched escaped simply because the wolves gave up quickly. In the majority of cases, the wolves turned back after chasing their prey less than half a mile. The vast majority of animals that wolves try to kill are simply too healthy and too fast to be caught. Examination of wolf kills indicates that most of the downed prey were diseased or weakened by age. The chief exceptions are healthy but very young animals who are simply too slow and weak to defend themselves.

A wolf's usual mode of hunting is to go out wandering, usually along well-worn trails. Except in the spring, when mother wolves go into underground dens to deliver and raise their young, wolves are almost continually on the prowl. If they happen to catch the scent of a prey animal, they often decide to test its vulnerability. The usual behavior is for the wolves first to stalk the quarry, creeping up on it as quietly as possible. The wolves move as close as they can—sometimes within twenty-five yards—until the prey animal discovers them. At this point, the prey must make a crucial decision. It can stand its ground and fend off its attackers as best it can or it can bolt and run, hoping to outdistance them. For many larger animals, such as moose, bison, mountain sheep, caribou, and the like, standing their ground usually discourages the wolves. The vast majority of animals taken by wolves are ones that run. For some reason wolves are reluctant to attack a stationary animal, preferring ones that are fleeing. One explanation is that an animal standing at bay can concentrate its attention on aiming kicks at lunging wolves whereas a running animal must watch where it is going, which leaves its flanks unguarded. An experienced wolf would presumably learn this difference.

The prey animal's decision is implemented the moment the wolves decide to rush it. If the animal stands its ground, the wolves move in to within a few feet, perhaps make a few feints, mill around a while, maybe take a sharp kick in the jaw or chest, and soon wander away. But if the animal opts to run, the wolves join the chase, usually stringing out in single file behind a leader. If the chase does not appreciably close the gap within ten or fifteen seconds, the wolves usually give up. If, on the other hand, the wolves catch up, the chase continues a little longer. It may go for a mile or two.

At this point, some chased animals decide to stand ground. The tactic almost invariably proves successful and the discouraged wolves lose interest before long. Other chased animals elect to continue running and, as if getting a second wind, most are able to put more distance between themselves and the wolves. Most outrun or outlast their pursuers, but a few, a distinct minority of even the few that the wolves pursue this far, are attacked and killed.

David Mech tallied all the wolf-moose encounters he witnessed during several winters of observation in the snowy woods of Isle Royale and came up with the following breakdown of the 131 occasions he saw wolves detect a moose. Eleven of the moose noticed the wolves and departed before the wolves even began stalking. Of the remaining moose that the wolves approached, twenty-four stood at bay without running and escaped harm. Of the ninety-six that ran when approached, forty-three got away before the wolves could catch up. The wolves caught up with fifty-three, of which twelve halted and stood at bay. But after a few moments of tension, these twelve were abandoned by the wolves. Forty-one moose continued running when the wolves caught up, traveling for a distance with wolves literally on their tails. Of these thirty-four eventually outlasted the wolves and escaped. Seven, however, were actually bitten and pulled down, as noted above, one of which escaped but six of which were killed and eaten.

Thus, of these 131 encounters Mech witnessed, just under 5 per cent resulted in a kill—hardly a stunning record for a "rapacious, killing machine" as some have characterized the wolf pack. The best you can say for the wolf is that most of the time he bungles the job or runs into a prey animal that isn't sick enough or weak enough to be killed.

The scientific picture of the wolf's lacks of persistence, viciousness, and success as a predator is hardly in accord with the image created by generations of fairy tales and writers' exaggerations. More in line with the truth is Mech's own account of the attacks he observed. One of the more dramatic is the incident in which a moose was attacked and wounded but not subdued. Mech watched from a light plane circling overhead:

"The moose had been browsing in the general vicinity of two others when ten members of the wolf pack approached. The wolves seemed to sense the moose when about three-eighths of a mile away, but did not appear to catch a direct scent until about 250 yards downwind of the moose. At this point the wolves suddenly charged straight toward the moose, which is the only time I saw them react this way while still so far from their prey. Two of the animals detected the wolves when just 25 yards away and they began running. The wolves followed them only a short distance and then saw the third moose which was closer. They immediately dashed the 50 feet to this animal and surrounded it. But the moose bolted away.

"The wolves instantly gave chase and soon five or six animals lunged at their quarry's hind legs, back and flanks. The moose continued on, dragging the clinging wolves until it fell. After only a few seconds it rose, then fell again. It rose again and fled toward a clump of trees, while the wolves continued their attacks. One wolf suddenly grabbed the moose by the nose, but the moose reached the stand of trees and the wolves released their holds. Beneath the trees the moose stood at bay, bleeding from the throat. Most of the pack lay down nearby, although a few wolves continued to harass the moose without actually biting it. The moose appeared strong and aggressive, kicking at the wolves when they approached its rear end.

"Darkness prevented further observations that day. The next morning the moose was still alive and in about the same spot. The bleeding had stopped but the moose walked stiffly. The wolves had abandoned the animal and had killed another, sixteen miles away."

Though the wolf is often portrayed as a bloodthirsty killer, slaying defenseless creatures at every turn without regard for its true food needs, the facts don't support this portrayal. The notion has been spread by hunters and other outdoorsmen who came upon carcasses only partially eaten and apparently abandoned. The scientific findings suggest just the opposite— by the time wolves have finished with a kill, little but splintered bones and scraps of furry hide is left. It has been found that if a prey animal is too large, a pack may be unable to finish it in a single sitting, though individual animals may put away fifteen to twenty pounds of meat. Though the wolves may go away for a few hours or even a day or so, they almost invariably return to gnaw and pick at the remains until little is left. Although wolves are their own scavengers, other animals such as foxes and ravens will steal from a wolf kill while the pack is away. To minimize losses to such freeloaders, an extraordinary adaptation of the wolf's digestive system allows huge quantities of food to be digested and the waste matter excreted in only two or three hours to make room for another meal as soon as possible. But after a day of repeated gorgings, the typical wolf may have to go from two to four days without food. During this time the digestive tract—adapted to a feast-or-famine existence—virtually shuts down.

Another remarkable adaptation of the wolf digestive tract is for the feeding of small pups. Unlike the lion, which usually lets its cubs scramble for the leftovers from a kill, the wolf regurgitates partially digested meat for pups. Whenever a pup wants food, it nuzzles the muzzle of an adult. If the older wolf has eaten recently, it readily, even eagerly, disgorges some of the food for the pup. The behavior, incidentally, is not limited to a pup's true parents. The burden of rearing the young is shared by the whole pack and any adult will readily feed a pup in its pack.

Of all the forms of wolf behavior, none is as well known as the howl, that melancholy sound that has frozen the blood of so many human beings who venture into the wilderness. To many it is a frightening sound, seemingly calculated to terrify and perhaps even to forecast imminent death for some individual that hears it. Again, the traditional view turns out to be wrong.

In the last three or four years popular attitudes toward the wolf howl have changed dramatically as the result of a concerted publicity campaign that has even included selling records of wolf howls. In 1971, for example, *Natural History* magazine issued a disc recorded in the wild and got a favorable review from Harold C. Schonberg, the acclaimed music critic of the New York *Times*.

Schonberg called the wolf "that sweet singer of beasts" and said of the wolves he heard that "the best of these virtuosos are capable of six- or seven-second phrases on one breath. Each phrase is a glissando swoop, up and down, like a very lonesome, sentimental fire siren with a soul. The best wolf singers start *pianissimo,* swell to a *messa di voce* to the sixth above, hold it sweetly and purely, then perhaps embellish to the upper partial before going down to a *pianissimo* and trailing off on an inconclusive microtonality near the tonic."

However the wolf howl rates in the technical terms of human music, there is some indication that it serves some of the same social functions. Far from any suggestion of evil or mournfulness, a group wolf howl suggests nothing so much as that disappearing phenomenon—the impromptu barroom songfest. One wolf will get up and let go with a long, swooping howl. The others in the pack immediately take notice. Others begin howling too, sometimes trotting over to where the lead howler is sitting or standing. As the solo expands to duo and trio, the whole pack may come over and join in, repeating whoop after howl until the air is filled with excitement. Tails wag enthusiastically and younger wolves move about friskily. Even the pups join in with yips and barks. Then, after a few moments, it all dies down and the wolves relax.

At other times the howl seems to have a more utilitarian function. When members of a pack are separated, one may begin howling. Soon the straying pack members trot back and assemble closely as if called together by the howl. Sometimes it seems as if wolves howl simply because they are lonely and want to establish communication with distant friends. An isolated wolf may start howling just to be answered by another howl from some distance away. The two wolves, or three, may howl back and forth several times as if conversing but without meeting. Quite possibly members of a pack know each other's voices and can recognize whether the answering howl is of a family member or of an alien. This guess is based on experiments with recorded howls, both genuine and imitations by human beings.

John B. Theberge, a professor of zoology at the University of Waterloo, in Ontario, who has conducted much of the howling research, believes the howl can contain significant information. "I discovered," Theberge wrote, in a scientific report, "that my three wolves could transfer information such as the identity of the wolf and perhaps its emotional state—by variation in the units of sound." Theberge's conclusions are based on his own analysis of individual howls recorded and displayed graphically by electronic devices showing that each wolf produces a characteristic combination of sounds. He found that wolves were able to distinguish between two similar howls that differed only in the intensity of a few subtle overtones—differences imperceptible to most human ears.

The popular interest in wolf howling has even led rangers in Ontario's Algonquin Park to take groups of visitors into the forest for nightly "howls." The people stop at various points, give out a few howls of their own and wait for the communicative wolves to reply, as they usually do.

Gradually, as the work of modern wolf researchers becomes better known, the howl of the wolf is coming to be understood not as a terrifying wail of death but as the song of a gentle, friendly animal that we are beginning to appreciate only now that we have banished him from our midst.

4

GORILLAS

King Kong, it seems, will never die. No matter how many air-
planes were sent to machine-gun him atop the Empire State
Building in Manhattan, he rose again to scale the World
Trade Center. And, no matter how many times animal behaviorists tell us that the gorilla is a shy, gentle creature, he will
continue to be thought of by some as the archetypal human-
oid monster.

More than any other species, gorillas lend themselves to a
caricaturization of what are thought to be the most beastly

qualities of men. Not man, men. Fictional gorillas are almost never female. Everything about them is masculine in extreme stereotype. They are big and hairy, with powerful bodies driven by primal instincts. The movie gorillas lumber along on two massive legs, smashing their way through civilization with one overriding drive—to find and rape a human female —it is *always* a human female who is his object. A beautiful but frail and refined Fay Wray. It's "Me Tarzan, you Jane" taken to the nth degree.

Gorillas make plausible movie monsters only if you don't know anything about gorillas. If you want to enjoy King Kong and all the others, never mind that naturalists and others who bothered to learn have been saying for decades that gorillas are not ferocious. Forget the fact that there have long been those who claimed that this introverted ape, who shuns human contact, is a lazy vegetarian who would rather munch a stalk of wild celery than attack people. And try to forget the fact, evident to anyone who sees even a zoo gorilla, that they walk on all fours and never get up on their hind legs except on very rare occasions and then only for a few seconds. Put all this out of your mind, for it will erode one of the most popular animal myths of this century.

Between 1908 and 1976 there were at least sixty commercially released American films depicting gorillas as ugly, brutish monsters filled with lust and violence. They include such titles as *The Doctor's Experiment, or Reversing Darwin's Theory* (1908), *Stark Mad* (1929), *Murders in the Rue Morgue* (1932), *The Monster Walked* (1932), *King Kong* (1933), *The Monster and the Girl* (1941), *Mighty Joe Young* (1949), *Bride of the Gorilla* (1951), *The Bride and the Beast* (1954), *The Beast That Killed Women* (1965), and the remake of *King Kong* in 1976. Most were dreadful films, Grade B and worse, but they nonetheless held millions of Americans, particularly young people, spellbound and doubtless helped shape some attitudes toward this wild species.

The films can be put into two categories: those in which the gorilla is represented as a real gorilla and those in which the gorilla is only a brutish façade, inside which is trapped a good and decent man. This latter theme harks back to the story of Beauty and the Beast, a fable common to many cultures for centuries, in which only a woman who recognizes the beast's innate goodness can transform him into a handsome prince or whatever.

King Kong, obviously, is one of these for inside the fifty-foot ape there beats a heart of pure goodness. In 1933 the beast pursues Fay Wray not to do her wrong but because he loves her. This befuddled Quasimodo in the concrete jungle of New York smashed buildings and cars and seems to those who do not know him to be bad but, folks, his heart is pure. His motives are misunderstood. It's as if the movie makers, all men of course, were trying to ask why women cannot see that no matter how mean their men seem to be, they truly love them and are good inside.

Of course, this is the stuff of tragedy. Paul Johnson, reviewing the 1933 *King Kong* in *The New Statesman*, said Kong was "more than a monster. He is a genuine character, a creature of intelligible rage, nobility of a kind and, above all, pathos. A prehistoric Lear in a sense."

Okay. But was Kong a gorilla? No, he was a human being and it's time we stopped maligning the apes.

The other category of gorilla movies—the ones purporting to show gorillas as gorillas doing gorilla things in Africa—have been no better. One of the earliest of such movies was

Ingagi, which was released in 1931 as a genuine documentary on gorillas. There had already been a dozen fictional gorilla movies in the preceding two decades. Presumably people were curious about the species and ready for some facts. *Ingagi* was said to have been highly successful at the box office. Unfortunately, the movie was anything but a documentary. It would probably not be remembered today if the Federal Trade Commission had not found its claims to being factual and authentic to be so blatantly false. Most of the scenes were shot not in Africa but in the Los Angeles zoo, with American black actors portraying Africans. Many of the "gorillas" were actors in furry costumes. Even the title, *Ingagi,* claimed by the filmmakers, Congo Pictures, to be an African word for "gorilla," turned out to be phony. As a result of an FTC order to stop misrepresenting the film, *Ingagi* was removed from commercial distribution.

How did it portray gorillas? One key scene had them coming out of the forest to carry away tender maidens offered as sacrifices by the African tribesmen.

Also in the gorilla-as-gorilla category one may include films in which real gorillas are filmed as part of the setting for a movie about human beings in Africa. A good example was *Mogambo,* the 1953 classic starring Clark Gable and Ava Gardner. The movie used real gorillas filmed in Africa. To get spectacular shots of gorillas charging, the film crew hired hundreds of Africans to locate a wild gorilla family and build a fence a mile and a half in circumference around it. The cameras were set up in a protected clearing near the fence while beaters gradually moved the fence, tightening the circle and, eventually, forcing the frightened gorillas toward the clearing. A human-shaped dummy was dangled near the cameras to provoke a charge and several gorillas responded spectacularly. The film was flown to a London studio and projected as a backdrop before which Gable and Gardner spoke their lines. The gorilla, again, was depicted as a ferocious, chest-beating monster, dangerous to man as well as to woman.

The myth of the ferocious gorilla is a relatively new one for Westerners since gorillas were not officially "discovered" (a word typically used to mean "seen by a white man and described in writing"; Africans, of course, have always known about them) until 1846, and it was not until the 1920s that scientists had much in the way of specimens to examine. It would be another forty years—the 1960s—before the scientific community, or anybody, for that matter, had much in the way of reliable behavioral observations to study. Until the reality of the gorilla was proven in 1846, tales of its existence circulated among Westerners in about the same way as tales of the Abominable Snowman or the Bigfoot of the American Northwest do today.

The first written suggestion that people had seen the gorilla, though not proof of that, is inscribed in Greek on tablets in the temple of Baal in Carthage which describe a voyage in 470 B.C. along the West African coast by Hanno, a Carthaginian explorer. The tablets, according to a Greek translation from the Punic, refer to the sailors finding, near Sierra Leone, an island "full of savage people, the greater part of whom were women, whose bodies were hairy, and whom our interpreters called 'gorillae.'"

Over the ensuing centuries there were scattered reports of huge, hairy beasts or wild men living in Africa. The first real proof of the existence of gorillas came in 1846 when two American missionaries, Thomas Staughton Savage and John Leighton Wilson, came upon some skulls in Gabon and sent them to British anatomists. Along with the skulls, Savage sent an account of gorilla behavior based, apparently, on stories from Africans:

"They are exceedingly ferocious, and always offensive in their habits, never running from man as does the Chimpanzee. It is said that when the male is first seen, he gives a terrific yell that resounds far and wide through the forest, something like 'Kh-ah!' prolonged and shrill. The females and young at the first cry quickly disappear; he then approaches the enemy in great fury, pouring out his cries in quick succession. The hunter awaits his approach with gun extended; if his aim is not sure, he permits the animal to grasp the barrel, and as he carries it to his mouth, he fires; should the gun fail to go off, the barrel is crushed between his teeth and the encounter soon proves fatal to the hunter."

One of the anatomists in England who received the skulls, Richard Owen, published his description of them and included in his report one of the stories he had heard about gorillas. "Negroes," Owen wrote, "when stealing through the shades of the tropical forest become sometimes aware of the proximity of one of these frightfully formidable apes by the sudden disappearance of one of their companions, who is hoisted up into the tree, uttering, perhaps, a short choking cry. In a few minutes he falls to the ground, a strangled corpse." It would be a century before scientists learned that gorillas spend almost all the daylight hours on the ground, going up into trees mainly to sleep and then never leaving their nests while it is dark.

Owen was impressed, as no doubt were many others who saw the skulls, by the size of the daggerlike canine teeth. They are truly huge, and in the thinking of that day such teeth were immediately assumed to be primarily weapons. We now know that long fangs serve at least as much for display, intended to frighten off attackers or to establish dominance within the group. In a mouth full of obviously vegetarian grinding teeth, like the gorilla's, the long canines are almost exclusively for display. But this is a recent discovery, and a century ago the gorilla's teeth seemed amply to confirm its reputation for ferocity.

While Owen was studying the bones and teeth, a French-born explorer named Paul Du Chaillu was tramping about Africa in the 1850s, hunting gorillas, and gathering material for what would become the single most decisive influence on Western thinking about gorillas. What Du Chaillu wrote about these gentle creatures would set the tone for a century of popular attitudes. In his 1861 book called *Explorations and Adventures in Equatorial Africa,* which is filled with boastful descriptions and obviously exaggerated tales of his heroic exploits, Du Chaillu tells of his attempt to shoot a gorilla. After what is said to be great difficulty, Du Chaillu closes in on the beast:

"And now truly he reminded me of nothing but some hellish dream creature—a being of that hideous order, half-man, half-beast, which we find pictured by old artists in some representations of the infernal regions. He advanced a few steps—then stopped to utter that hideous roar again—advanced again, and finally stopped when at a distance of about six yards from us. And here, just as he began another of his roars, beating his chest in rage, we fired, and killed him."

Du Chaillu omits details of just how the gorilla "advanced," but it almost surely was on four legs, rearing up on the hind legs only to beat the chest. We now know that if Du Chaillu had simply stood there and watched, the gorilla would probably have decided that he could not scare away the intruder and the beast would merely have turned and scampered away into the forest. People who study wild gorillas today almost never take along firearms. They have seen the loud and, to the uninitiated, frightening "threat displays" of the male gorillas with all its roarings, flashings of fangs, resonant chest beatings, and rushes that invariably stop short of actual attack. It is, to be sure, an impressive display, especially when executed by a six-foot-tall, four-hundred-pound gorilla with an eight-foot arm span. (The bloated six-hundred-pound zoo gorillas, lacking enough exercise and often overfed, are grotesquely fatter than their free-living relatives.) The display, we now know, is all a routine ceremony, an innate ritual that gorillas perform when they get excited.

In the 1960s both of the two principal modern gorilla observers, whose findings have revolutionized the ape's image, George Schaller, who "did" gorillas before he "did" lions, and Dian Fossey, another young American zoologist, emphasize the almost total absence of truly aggressive or hostile behavior on the part of gorillas.

Fossey, who had been studying gorillas in East Africa for the last ten years, says that in the first three thousand hours of observation she recorded "only a few minutes" of aggressive behavior. She recounts one incident that could have been described in the best Du Chaillu lurid prose. Five huge males charged her simultaneously, all roaring—no doubt "hideously," as Du Chaillu would have put it—and quickly closing in on the young woman, who had no weapons. When the biggest gorilla was barely a yard away, Fossey spread her arms and shouted, "Whoa!" All five big male gorillas stopped and eventually ambled away—sheepishly, one is tempted to add.

Schaller describes, in his book *The Year of the Gorilla* (1964), an incident in which two gorillas used much the same routine on each other. It occurred when two groups of gorillas happened to meet and a male from one group confronted a male from the other. The incident began when they were about twenty feet apart and one gorilla began to get excited.

"He hooted softly," Schaller wrote, "and with increasing tempo until the sound slurred into a harsh growl; he beat his chest, wheeled about, lumbered up a log and with a forward lunge jumped down to land with a crash. As a finale he gave the ground a hollow thump with the palm of his hand. The Climber [as Schaller named him] walked rapidly toward the other male and the two stared into each other's eyes, their faces a foot apart."

Long before this point Du Chaillu, had he been confronting one of the apes, and every other gun-bearing human being probably would have fired, convinced that the alternative was certain death.

"These giants of the forest," Schaller continued his narrative, "each with the strength of several men, were settling their differences, whatever they were, not by fighting but rather by trying to stare each other down. They stared at each other threateningly for from twenty to thirty seconds, but neither gave in, and they parted."

This sort of display has been observed a number of times, in the wild and in captivity, and it follows a fairly predictable pattern. It begins with some hooting. Then the gorilla gets up and starts grabbing branches or tussocks of grass and flailing them about wildly. Sometimes the gorilla scampers about, grabbing anything in reach or thwacking it with a flailing arm. Anyone in the way could get hurt inadvertently. The next stage is chest beating, which females and even babies do. The last act is a short-distance run that may end with a slap at the ground. If the thing doing all this is a four-hundred-pound gorilla strong enough to rip down branches the size of a person's arm and the run is toward you, you are supposed to get scared. If you don't, the gorilla concludes—to be unabashedly anthropomorphic—that you have called its bluff and it goes away.

Through the research of the Austrian zoologist Konrad Lorenz and the Netherlands-born British zoologist Nikolaas Tinbergen we now know that this type of behavior is rather common among animals. It results when the beast is faced with a choice of either confronting an intruder or running away. Lorenz, who is with the Max Planck Institute in Germany, and Tinbergen, a professor at Oxford University in England, theorize that the tension between these alternatives creates such a state of anxiety and unease—probably from the sudden flow of adrenalin—that the animal looks for ways to displace its tensions. It has to do something to drain away the nervous energy. Many animals, such as bulls, do it by pawing at the ground. Some start feeding. Some, such as antelopes, urinate or defecate. Some, such as elephants and chimpanzees, rip up grass or branches and throw them about. Some start pacing or swaying nervously. Even human beings facing a difficult choice relieve tensions by similar means. People drum their fingers or scratch or tap their feet. A quarreling couple, for example, may, to avoid physical attack on each other or total retreat, start throwing objects or slamming doors or kicking cats—all as a way of draining off energy without doing harm to the adversary or the ego. As any marriage counselor or wildlife biologist will tell you, the truly dangerous situation arises when the aggressor makes no display and, his mind already made up to attack, comes quietly and directly toward his target.

In other words, the gorillas that seemed so ferocious and evil to their "discoverers" were simply having a hard time making up their minds on how they felt about being discovered. Schaller has reported instances when he startled gorillas who then put on their best display and then stopped to peer inquisitively at Schaller as if to see whether it had any effect.

Perhaps the most widely held belief about the gorilla is its sexual prowess. A good proportion of the gorilla movies rely on this theme, the most clichéd scene being one in which the hairy brutes carry some scantily clad woman about the city or jungle. The appeal of such a scene to a human audience is obvious. Women might shiver with fear but, believing that gorillas are highly sexed and well endowed, women might also find the scene vicariously exciting. Men, sharing the myth, might fantasize themselves in the role of the gorilla.

The story of the gorilla's great sexuality is an old one, antedating even the official discovery of the creature. For example, in the unexpurgated *Arabian Nights,* a classic that is at least several centuries old, Scheherazade tells the story of Ladite, a young noble woman who was having an affair with a gorilla, meeting him every night in the palace basement. There are many versions of the story and some are extraordinarily explicit even by today's standards, giving many details of the elaborate love games enjoyed by the woman and the male gorilla. Eventually someone kills the gorilla and learns from the woman the history of her relationship. From her youth, she said, she had been taken care of in the palace by a large black slave who initiated her to sex. The story also illustrates the antiquity of the myth of the sexual prowess of black men, for, Scheherazade says, when the slave died, the woman was advised that the only creature capable of rivaling her human lover was a gorilla. Ladite, the woman, promptly bought a gorilla and took up with him.

More recent literature, such as *Tarzan of the Apes* (1914), by Edgar Rice Burroughs, has also reinforced the idea of ape sexuality. Tarzan meets Jane, as fans will recall, after rescuing her from a giant ape who, intent on having her for his own, slings her over his shoulder, and sets off through the trees. Tarzan, having been raised by the apes, has no trouble pursuing and saving the fair damsel.

The various King Kong films have, of course, carried on the theme. At the end of the 1933 version, when Kong topples from the skyscraper, itself a phallic symbol, there is discussion of what really spelled the ape's doom, and Denham, the movie producer in the 1933 film, says, "'twas Beauty killed the Beast."

Real movie producers must also think like that, for, a few years ago, when John Huston returned from Africa where he was making a picture called *The Roots of Heaven* (1958) with a large number of female stars, he observed that not one of the women was ever molested by a gorilla.

No wonder. Gorillas, serious observers have found, are about the least sexually active species of primate. Geoffrey H. Bourne, director of the famous Yerkes Regional Primate Research Center in Atlanta and a world authority on apes, says, "Man may be the sexiest of all primates, and the gorilla may well be the least sexy. This certainly must come as a surprise to those who believed the gorilla to be the supersexed stud of the animal kingdom, but experts give a very low score to the gorilla as a lover." In his book on gorillas, *The Gentle Giants* (1975), written in collaboration with Maury Cohen, Bourne cites a number of authorities to the effect that an adult male gorilla's penis is extremely small, the average length being around two inches when erect.

What's more, Bourne says, male gorillas aren't at all attracted to human females, at least not sexually. He reports an incident at a Swiss zoo where a young woman employee accidentally locked herself into a cage with a solitary adult male gorilla just at the zoo's closing time. "The animal," Bourne wrote in his book, "was very happy to have her company and placed his arms around her and took her over to the side of the cage, where he lay down and slept with his arms around her all night. He made no attempt at any sexual play of any sort." The woman could not extricate herself during the night, and in the morning, when the ape released her, she walked away shaken but totally unharmed. The animal, imprisoned in his cell, had just been lonely.

Even among themselves in the wild, gorillas apparently aren't terribly interested in sex. Over his fifteen months of observing the groups of gorillas comprising 169 individuals, George Schaller witnessed only two copulations. From his own observations and from accounts from zoos where gorillas have been bred, Schaller concluded that gorillas become interested in sex only for three or four days during each month when the female is ovulating. At such times, the female generally does the soliciting of the male of her choice—usually one of the higher-ranking males in the family hierarchy. A gorilla family, varying in size from three or four animals to one or two dozen, usually includes one dominant male, several lower-ranking adult males, several adult females, and a number of juveniles. Since the females do not ovulate when they are pregnant or lactating, and since most mature females are usually doing one or the other, male gorillas may go for as much as a year without sexual intercourse. While one might suppose this makes the males feel extremely deprived and eager for sex, Schaller saw no such indications among the males. They did not express the slightest visible sexual interest in females except and until the females were ovulating.

Thus, it appears that gorillas, far from being the lascivious, rapacious beasts of fiction, are quite restrained in their sex lives, even obeying the orthodox dictum that sexual intercourse should be only indulged in for procreation. For human beings, of course, it has many other functions, and in discussing those who abhor modern human sexual freedom, Bourne says, "Apparently these individuals believe that in sexual relations we should behave like apes."

Actually, that is not quite fair to gorillas, for when they finally do get around to mating, they may do it in almost any position and, unlike most animals, they seem to display a bit of tenderness and affection before and after coitus. Observations in zoos indicate that gorillas sometimes play during copulation and, after climaxing, there is sometimes a bit of mutual fondling and petting.

By discarding the myth of the gorilla's innate viciousness, George Schaller was able to approach groups of them, coming within their view and following them about in their daily travels, to learn how gorillas really live. His work has been published in two forms, his popularly written book, *The Year of the Gorilla,* mentioned above, and an earlier scientific monograph, *The Mountain Gorilla* (1963).

Schaller learned that the gorilla's day is hardly one of charging hunters or chasing women or running amok in civilization. It is a gentle, unhurried, even monotonous and humdrum existence that, for all the gorilla's strength and energy, proceeds as slowly as the sun crossing the sky. The gorilla's day begins around seven in the morning, after the sun has risen. (Near the equator sunrise and sunset come at about the same time—6 A.M. and 6 P.M.—all year round.) Having spent the night in individual nests either on the ground or up in the trees, the band of gorillas rouses slowly. The nests are simple one-night affairs made by the gorilla staying in one spot and pulling toward itself all the nearby branches of trees, or undergrowth when nests are on the ground, breaking them down toward the center of the nest. The best nests are in trees, and the broken branches, still attached at the breaks, form a kind of platform.

Once they awaken, the gorillas begin feeding in the immediate area, pulling up wild celery or bamboo shoots, plucking leaves, and chewing on stems. They are strictly vegetarian. All the animals, from the dominant big male to the subordinate younger males, if any, and the females and infants, amble about on the ground on all four legs until 10 A.M. Then, as the air is warming up, they take a siesta, sometimes sleeping, sometimes just sitting and grooming one another while the little ones play.

By midafternoon the siesta is over, and the entire family, at a signal from the dominant male, assembles in single file and moves off to another feeding ground. Usually this is no more than half a mile away. There they stop and feed again for a few hours until about dusk. By 6 P.M., as the sun is going down, the big male starts to make a new nest. This is the signal for all the others to do the same. Nests are not used more than once, a new one being constructed each night.

By dark all the gorillas are in their nests and they do not get out again until the next morning. If they must defecate, they do it in the nest. Gorillas sleep twelve hours a night and siesta for four hours or more in the middle of each day. Schaller estimates that 80 to 90 per cent of a gorilla's waking hours are spent on the ground. Furthermore, much of the night is spent on the ground for that is where they often make their nests. That they bother to make a nest on the ground is curious, for it appears to serve no purpose. In the trees, the nest forms a platform with a rim to keep the animals from falling. On the ground, however, there is not enough nest bottom to keep them off the damp ground. And, if the gorillas choose to build the nest on a hillside, the rim is not usually enough to keep them from rolling out. Schaller says gorillas sometimes wake up in the morning to find they have rolled out of their nests and down the hill ten or twenty feet.

The nest in the trees is virtually the gorilla's only strong link with its arboreal evolutionary past, and the fact that the nests are often only flimsy vestigial affairs on the ground suggests that this link is fading. When gorillas climb into the trees, they are often clumsy and awkward—hardly the picture of a creature superbly adapted to the arboreal way of life.

"Gorillas are not agile," says Schaller, "and they are relatively poor judges of what branches can support their weight. Several times I have seen gorillas step on a twig which snapped and the animal was only able to save itself from a serious tumble with a firm handhold. Once, in the distant past, gorillas were probably arboreal, but now with their huge size and swaying belly they seem anachronistic among the boughs." (Climbing up the Empire State Building or the World Trade Center towers wouldn't be any easier, either.)

Schaller completed his fifteen months of studies in Uganda in 1970 and went on to study Indian tigers and African lions. In 1967 Dian Fossey began her studies of gorillas in the Congo Republic (since 1971, Zaïre). Because of local political problems, she had to transfer to nearby Rwanda after only six months. She has remained there, off and on, ever since and is today regarded as the pre-eminent authority on the behavior of wild gorillas. Unfortunately, she has not published much of her research except in her doctoral thesis, a few brief scientific reports, and articles for the *National Geographic*. Therefore, although she has reportedly developed some interpretations of gorilla behavior that differ somewhat from Schaller's, the details are not fully available.

Schaller exploded most of the popular myths about gorillas, but Fossey has gone further by establishing communication with these comparatively intelligent creatures. Where Schaller merely observed the animals, Fossey has tried to move in with them by imitating gorilla behavior and becoming accepted as a member of their family groups. She has succeeded beyond most zoologists' expectations and her achievement should drive the final nail in the coffin of stereotype about violent, lustful gorillas.

One of the most revealing incidents that Fossey has reported occurred one day after three years in the field, when she had yet to become fully accepted by the beasts. On that day a male gorilla suddenly approached Fossey to within a few feet and began beating his chest and strutting in the typical four-legged stiff walk that gorillas do to make themselves look big and powerful. She knew that it was not threatening behavior and interpreted it as an effort to relate to her. To reply, Fossey imitated certain gestures she thought gorillas used to show friendly intent. Fossey scratched her head noisily and, with her other arm, scratched under the first arm. The gorilla stopped, stared a moment and then imitated her perfectly. Could the huge beast be trying to communicate? Fossey slowly reached out a hand and turned her palm up. The gorilla stiffened for a moment, apparently uncertain of how to respond. Then he reached one hand forward and gently touched Fossey's hand. Then the gorilla scampered away, beating his chest, and rejoined his group some distance away. Fossey began to cry. "To the best of my knowledge," she said, "this is the first time a wild gorilla has ever come so close to holding hands with a human being."

Since then, Fossey has developed still closer bonds to the gorillas and frequently plays with them in a form of wrestling. She can sit side by side with a gorilla, just staring off into the distance, each with an arm around the other.

Fossey has also contributed at least one bit of evidence that should be allowed to temper the emerging popular view of gorillas as totally pacific beasts: they do occasionally become violent toward one another. Fossey calculates that 15 per cent of their physical interactions are violent. That means 85 per cent are nonviolent and that gorillas are, on the whole, gentle. But imagine how that level of violence toward one's own kind would translate into human terms. If every eighth person with whom we interacted physically hit or hurt us in some way, people would have to be considered far more violent than they are. Most human beings go for weeks or months, engaging in hundreds of physical or near-physical contacts with other human beings, never once being physically hurt. Man's alleged propensity for violence is dealt with more fully in Chapter 12.

So, gorillas are not the vicious beasts of our myths and they are not totally cherubic either. They can occasionally be quite nasty. Fossey has observed three instances in which adult male gorillas deliberately killed baby gorillas, and she has indirect evidence of three other such cases. Gorillas, like all other animals viewed in an anthropomorphic sense, have their good traits and their bad traits. A mature appreciation of gorillas must accept both.

Western man's relationship to the gorilla is, as we have seen, of rather recent origin. Though they were discovered more than a century ago, Americans and Europeans have had little on which to base their understanding of these huge apes until the last fifty years. As late as 1915 no zoo had kept adult gorillas. Until then, animal collectors could capture only babies when they shot the adults in a family. Unfamiliar with gorilla diet and environmental needs, zoo keepers could only watch dozens of young gorillas dwindle, physically and mentally, in their cages. When one died—it usually had no companions— the zoo bought another one. Eventually zoos learned how to care for captive gorillas and raise them into adulthood. The first successful breeding of gorillas in captivity occurred in 1956 at the Columbus, Ohio, zoo. Thus it is only in the last twenty years that Westerners have had an opportunity to gain anything like a reasonable idea of what gorillas are. Unfortunately, however, too few zoos exhibit their gorillas in family groups. Too often they are separated into individual cages. George Schaller's two pioneering books on gorillas— the first truly reliable accounts—were not published until 1963 and 1964. Dian Fossey, who apparently has vastly extended Schaller's studies, has yet to publish her magnum opus.

Because of the exploding human populations in Africa and the rapid growth of agriculture and human settlements, the mountain gorilla's range is being rapidly compressed. Although very little is known of the exact status of gorilla populations—Dian Fossey is devoting much of her current work to studying this—it appears that the mountain gorilla survives in only a few dozen pockets of central Africa, mostly on mountain slopes totally surrounded by cultivation. It is estimated that somewhere between five thousand and fifteen thousand remain. The lowland gorilla, which is smaller, less furry, and lives mostly in western Africa, is believed to be somewhat more numerous. Both races of gorilla, however, are severely reduced in numbers from only a few decades ago. Fossey believes that without a concerted effort to protect the mountain gorilla's habitat, the animal will become extinct in the wild within the next two or three decades.

Because the needed conservation action is unlikely to come on a large scale very soon, it is probable that the mountain gorilla will be reduced to perhaps only a half dozen protected parks with only a few hundred surviving animals. Whether that number, whatever it turns out to be when protection becomes effective, will be sufficient to guard against total extinction from fatal epidemics or other natural disasters remains to be seen.

Perhaps by the time the gorilla makes its exit, people will have come to appreciate it for what it was.

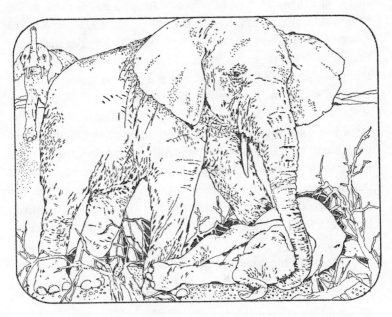

5

ELEPHANTS

The circus! Ringling Brothers and Barnum & Bailey! The Big Top! Cracking his whip and wheeling on his heel, the ringmaster, black boots gleaming, booms into the microphone: *"Ladeeeeees and gentlemen!* Here they come, direct from the jungles of Darkest Africa! A plethora of ponderous pachyderms!"

And, sure enough, the elephants lumber in, single file, a chorus girl sitting on the head of the lead and biggest elephant while half a dozen other elephants trot dutifully behind, each trunk daintily grasping the tail ahead. Somebody carrying a long stick with a hook on the end keeps yanking at the elephants but you assure yourself that they wouldn't really hurt them.

The elephants go around and around in a circle inside the center ring. One does a handstand. Another stands in the middle while the others rise up on their hindlegs and lean on their front feet against it. The chorus girl swings down to the ground and lies flat in the sawdust while an elephant hovers its foot inches above her head. The elephants—that fellow keeps yanking them with the hook—turn around several times as the ringmaster announces that they are dancing to the music.

In a little while it is over. The elephants plod away and many people think they've seen real elephants. Millions of people, witnessing the same kind of act, think they've seen real elephants. The beasts might as well have been trained seals or, for that matter, trained fleas, for they did nothing really elephantlike. The amusement was not in displaying elephants for what they are but in creating a spectacle by making a three-ton creature prance like an acrobat.

Now, let's go to the other extreme. It's a 1950s movie and we're in Africa, in the elegant villa of a white man who, to prove his theory of the secret graveyard of the elephants, has been searching for the legendary place they all go to die. The handsome hero, maybe Farley Granger, tells his beautiful wife, say, a young Elizabeth Taylor, that he's doing it out of a noble spirit of quest but we know he's been talking to Sydney Greenstreet about all the ivory that must be hidden there.

One day, after weeks on safari, Granger returns home to report to Greenstreet he has finally found the secret valley of the elephants and tomorrow will hire a thousand porters to carry the ivory out. That night at the villa the warm tropical air is shattered by the trumpeting of hundreds of elephants who break out of the forest, stampede through a banana plantation, and come crashing into the house seeking revenge for Granger's violating the elephants' graveyard. Granger leaps out of bed, grabs his rifle, and steps onto the veranda in time to be crushed by an angry bull. The elephants rampage through the African village, ripping up huts and scattering children and chickens. Back at the villa the leader bull elephant finds Elizabeth Taylor, grabs her in his trunk, and carries her off into the forest. The secret of the graveyard is safe once more.

Those weren't elephants either. They were ponderous pachyderms trained by Hollywood to act like people.

Incredible as it may seem, nobody from the industrialized world learned in detail how elephants behaved in their everyday life until recently. For only then did skilled animal behavior researchers actually go out into elephant country and sit still with their eyes open while elephants lived their lives over a period of years.

What these scientists found was that elephants are far more intelligent and behaviorally complex creatures than Ringling Brothers or Hollywood ever imagined. Elephants have one of the most advanced and enduring social structures in the mammalian world and can adapt their life styles to almost as wide a variety of habitats as man. They have been found in every environment from grassy plains to dense forests to mountain slopes almost up to the snow line. They have distinct individual personalities and pass on knowledge from one generation to the next.

Elephants live in stable family units ruled by a matriarch who is the mother, grandmother, sister, or aunt of every other member of the family, and she may remain the head of that intact family for several decades—typically until death, which may come in an elephant's forties or fifties. When young males (bulls) approach adulthood, they are pushed out of the family and go off to live a more or less solitary existence. Contrary to a common opinion, the bull elephant is not the protector or leader of a family. Adult bulls live on their own, occasionally traveling in loose association with other bulls, occasionally checking out the cows in some nearby family unit to see whether any are in heat.

In fact, if danger threatens a family unit while a bull happens to be nearby, the bull is often the first to run. The family's matriarch, on the other hand, is invariably the protector who faces the danger, perhaps charging it, while the other elephants form a defensive ring with the bigger elephants on the perimeter facing out and the smaller ones safely in the middle or, if they are babies, under some bigger elephant's body. The matriarch is the one who charges and in retreat always places herself between the family and the threat.

For decades, many professional hunters, asserting that no one could know an animal as well as they, believed that elephant families were led by old bulls, some over a century old they would claim, citing the heavily wrinkled bodies. In fact, an elephant's life span is less than that of human beings. Few elephants live beyond their fifties. The reason is their teeth. In addition to the tusks, which are specialized teeth, an elephant has, at any one time, only four functional teeth—one on each side of the upper and lower jaws. They are long molars that grind almost constantly every day. When the first set of teeth wear down, they fall out and a second set comes in from under the first. Elephants have exactly six sets of teeth, each appearing in its turn, and when the sixth set is worn down, the elephant must find enough soft vegetation that needs no chewing or starve to death. Eventually all older elephants that are not killed by people succumb to malnutrition and related diseases.

In retrospect, it is easy to see why the macho mentality of many hunters would have made it difficult to accept the idea that anything other than a big, brawny male could lead an elephant family.

Today, thanks to the work of a handful of researchers whose findings date mostly from the mid-1960s, we know better. The scientist who has made perhaps the most thorough and sustained observations of elephant behavior is Iain Douglas-Hamilton. For four and a half years Douglas-Hamilton lived in Tanzania's Lake Manyara National Park studying what is believed to be the densest elephant population in Africa. Douglas-Hamilton, a student of Tinbergen at Oxford, came to recognize individually more than four hundred of the park's five hundred or so elephants. He observed one particular kinship group of about fifty elephants on 314 days over a four-year period. His doctoral thesis at Oxford, "On the Ecology and Behaviour of the African Elephant" and the more popular book he wrote with his wife, Oria, *Among the Elephants* (1975), offer what is probably the best comprehensive view of African elephants in print.

Each elephant, Douglas-Hamilton found, has a distinctive personality. In one family unit in Manyara, for example, there were a pair of adult females (cows) who were virtually inseparable. They did everything together. One, however, was a very inquisitive beast who seemed fascinated by Douglas-Hamilton and who often approached him closely and over a period of years formed a kind of friendship with the human being that allowed them to touch each other without fear. Her elephant friend, however, who was nearly always around during such encounters, always hung back, nervous and afraid. She never warmed up to people.

Modern research on elephants has helped explode one of the most persistent views about wild animals in general—that the young are born with all their instincts, ready to face the dangers of life in the wild. This view assumes that man is the only species in which the adults take much of a role, other than feeding, in raising the young. Hence, it is often asserted that the human family with its long-lasting bonds between parent and child is unique. While it is certainly true of many animal species that parents have few responsibilities toward the young and that family ties are weak or absent, it is most definitely not so with the more intelligent species such as apes and elephants where families are stable and close throughout lifetimes.

Anthropologists generally agree that the long years during which the human child is dependent on its parents are linked to the institution of the stable family. The same is undoubtedly true of the elephants.

Biologists who study the evolution and nature of intelligence have found a close correlation between the length of youthful dependency and the level of intelligence in the species. In animals such as the birds or reptiles or fish that rely more on instinct (immutable behavior patterns programmed into the brain by heredity), there is little to learn after birth. Such a species' repertoire of behaviors is already "wired" into the brain. Such systems work fine as long as the habitat of the species remains stable and offers no new challenge for which the brain is not prepared. But, if the species is to be adaptable to changing environmental conditions, a fixed set of behaviors can become a prison. To survive under such conditions, a species needs a generalized intelligence, capable of learning whatever is useful in any given time and place. Evolution's standard method of achieving this versatility is by exposing the young to the world when their brains are still quite immature and letting them develop not in the fixed environment of the uterus but in the constantly changing outside world.

Elephants are born with their brains only one-third adult size. There is enough brain to operate the internal organs but not enough, for example, to manipulate the trunk. Newborn elephants, called calves, are constantly tripping over their trunks. It is months or years before they learn, for example, to drink with their trunks; until then they kneel awkwardly and submerge half their faces to get water to their mouths.

"The older calves that I studied," Douglas-Hamilton wrote in *Among the Elephants,* "showed that for at least the first ten years of their life, elephants continue to be nurtured by the love and protection of their families. Even when the next sibling is born, the older calf still receives plenty of affection from his mother which goes on until adolescence and in some cases long beyond."

It is evident from Douglas-Hamilton's studies and those of other researchers that there are not only individual differences among elephants but differences between groups that can only be called cultural differences. For example, in the northern end of the long Lake Manyara National Park, the elephants are generally placid in the presence of human beings. In the park's southern end, on the other hand, the elephants were much more excitable and quicker to charge intruders. In this part of the park there lived four cows, nearly always found together, whom Douglas-Hamilton calls "the dreaded Torone sisters." Whereas most elephant charges are bluffs that stop short of actual contact with the chargee, the Torone sisters would attack in earnest, plowing their tusks into Land-Rover doors and radiators, pushing vehicles around, bashing in roofs. A third group of elephants, who preferred the more heavily wooded parts of the park, shunned contact with people altogether, retreating into the forest at the slightest approach of a person. Why should these groups behave differently, especially since all live in the same protected environment of a national park and have not been threatened by hunters in many years? Perhaps—and this is one popular idea about elephants that appears to be true—they remember the way they were treated many years before. There is even evidence that traditional behavior patterns among a group of elephants may be passed on from one generation to the next by some form of instruction.

Take, for example, the case of the elephants of Addo Elephant National Park in South Africa. In 1919, in the area where the park exists now, there was a forest harboring an estimated one hundred forty elephants that frequently came out to raid the neighboring citrus plantations. The farmers hired a hunter to exterminate the huge pests. The modern elephant-culling method is for a whole family group to be shot in a matter of seconds, but the hunter who got the Addo job in 1919 shot the elephants one by one, as targets of opportunity, over a period of months. The surviving elephants, hearing or even seeing the shots and knowing that one of their family had been killed, were no doubt deeply affected. In those days little was known of the cohesiveness of elephant families. After a year of random shooting, no more than two or three dozen elephants remained and the hunter sought to finish them off with one last shooting campaign. However, the elephants had grown so wary of the hunter, his odor, and the sound of his rifle that at the first sign of his approach they would disappear into the thickest part of the forest and not come out until after dark. On some occasions when the hunter pursued them, they would counterattack and he would only narrowly escape with his life. Eventually the hunter gave up.

By 1930 the government of what was then the Union of South Africa, having witnessed the extermination of nearly all its wildlife through hunters and loss of habitat, began to change its policies and established a sanctuary for the Addo survivors. An elephant-proof fence was erected around the area and the animals were no longer hunted. Yet today, more than half a century after the hunting campaign, the Addo elephants are still known as the most dangerous elephants in Africa. Unlike most elephants elsewhere, they remain nocturnal and readily charge any human being who comes near. Most, if not all, of the elephant survivors of the extermination attempt would long since have died. Yet their descendants still maintain the tradition of wariness in the presence of man.

The war against elephants, who unquestionably have destroyed the livelihoods of countless thousands of African farmers by stealing and trampling crops, continues today. The human populations of East, Central, and Southern Africa, where the major elephant concentrations remain, are growing rapidly and the demand for food is steadily converting more and more wilderness, including elephant country, into farm and ranch land. To escape the threats of both angry farmers and ivory hunters, elephants are retreating to the safety of the national parks and game reserves. A prime example of this phenomenon involves the elephants in and around Kenya's Tsavo National Park. The situation there is worth examining in detail as a classic example of how elephant behavior, human sentiments about wildlife, and government efforts toward scientifically based wildlife management can become entangled. Probably no other controversy has done more to divide the ranks of conservationists and to cripple ecological research in East Africa than that involving the elephants of Tsavo.

Because of pressures outside the huge eight-thousand-square-mile Tsavo National Park, thousands of elephants have been crowding into the santuary, swelling the population to somewhere between twenty and thirty thousand, making it one of the largest concentrations of elephants in the world today.

Because elephants eat so much and push down so many trees in the course of their activities, a large concentration can devastate an area. Tsavo, once a lush land of dense bush and trees, is today a sad landscape of bare ground and struggling grasses, strewn with the skeletons of downed trees. Because the park receives very little rain, some scientists have said this destruction could eventually turn Tsavo into a desert, providing so little food that most of the elephants would eventually starve to death. The long-term result could be a "population crash" that would wipe out elephants in the very place set aside to protect them.

In 1970 and 1971 a drought reduced the already damaged Tsavo vegetation so severely that between five and six thousand elephants died of malnutrition. Since then, still more elephants have entered the park, seeking refuge from hunters and expanding agriculture. The population may now be as high as before the drought. Further droughts and massive die-offs are almost certain in coming months and years.

When the Tsavo Research Project team recommended shooting three thousand Tsavo elephants for research purposes (detailed autopsies can reveal much about a species' state of health) and indicated that it might be necessary to shoot many more to bring the population back into balance with the environment, many conservationists recoiled in horror. Killing the animals they were trying to protect hardly fit the traditional conservation ethic.

Richard Laws, the British director of the project, was recognized as one of the top men in his field—large-mammal ecology—but the outcry at the recommendation forced him to quit his post. The park warden, a former officer of the Kenya regiment of British colonial days, who bottle-fed orphaned baby elephants, vowed never to permit such a step. The rumor became widespread that the whole proposal was a money-making scheme by the scientists in conspiracy with a commercial concern that was to do the shooting. Indeed, it has been estimated that selling the ivory and skins from the three thousand elephants to be shot would have earned a profit of over $1.5 million.

The proposal was so bitterly opposed and so widely misunderstood that it turned many of East Africa's park wardens, some of whom were retired British military men with little more than a hunter's myth-filled knowledge of ecology, into active opponents of research aimed at better park management. Outside Africa many conservationists have come to doubt the motives of scientists in Africa. The Tsavo Research Project, which was slated to become a major research center on large-mammal ecology, is today a small, struggling organization unable even to observe properly the extraordinary phenomenon taking place on its doorstep.

When it was founded in 1966, the Tsavo Research Project received a $220,000 grant from the Ford Foundation to cover three years. When the grant expired in 1969, the financially pressed Kenya National Parks took the project over but only assigned it some $42,000 a year. On such a budget, many studies had to be abandoned, and others were kept going because of donations from various private conservation groups. Most of this money comes from the United States.

However, while in the United States teams of scientists can concentrate on the subtle changes in individual ponds, at Tsavo one of the largest and most spectacular ecological transformations in the world must be largely ignored for lack of funds and researchers. The handful of scientists that do remain there are unable to determine accurately even so elemental a fact as the number of elephants in the park, although methods for doing this exist.

If little is being done to understand the phenomenon, nothing is being done to control or manage it. It may, in fact, be too late. Today vast areas of the park, which is just a bit larger than New Jersey, have become what one scientist called "elephant slums."

Philip Glover, who succeeded Laws as director of the Tsavo Research Project, stood outside his laboratory in the park one morning in 1974, squinted into the distance, and told me: "All that out there used to be thick bush. Now look at it. Just a year or so ago this area right around the building was so thick you couldn't see beyond. Now they [the elephants] are coming right up to the houses."

Dead and twisted trees lay all about. The ground was a patchwork of dry grass and dusty soil. In the distance a family of elephants grazed, ripping up tussocks of standing hay with their trunks and stuffing them into their mouths. A full-grown elephant needs about seven hundred pounds of grass, leaves, and twigs a day; how the Tsavo elephants can get it from such a meager landscape is hard to see. Except in the brief blooms of green following the rainy season, they probably don't any more. One of the things Laws's proposed autopsies would have revealed is the state of nutrition of the elephants, some of which would have been shot during the time when there was plenty of food and others of which would have been taken during the dry seasons.

"It takes a lot of patience to stand by and do nothing," Glover said. There was sadness in his voice, and a touch of anger. "But it's a natural phenomenon," he quickly added, as if to check an emotion that shouldn't fog a scientist's view.

If the trees cannot survive the elephants, they will be replaced by coarse grasses that can. If, then, the land cannot support the elephants, they too will die, and something else will replace them. Few ecosystems are constant and unchanging and there is really no such thing as a "pristine" state of nature. Still, ecosystems rarely change as rapidly as that of Tsavo, nor in a spectacle as saddening. Much of Tsavo looks as if a war had just ended there. It is a ravaged land. The elephants shuffling about, scrounging bits of grass under fallen trees, are both the war's orphans, and its combatants.

Problems of high elephant density are developing in several parks around Africa, all more or less for the same reasons—increased competition from people. Because of this factor and increased illegal hunting for ivory, Africa's elephant population as a whole is dwindling, some say rapidly. Although there are said to be over five million elephants in Africa, a century ago there were probably twice as many.

Because so many elephants are now crowded into the parks, the beasts are showing signs of ill health—the same physiological effects of overcrowding that are seen in artificially crowded populations of animals such as laboratory rats. Fertility is declining and infant mortality is increasing. The calving interval, normally about four years (the gestation period is twenty-two months), has lengthened to as much as nine years in some populations. In the first four years of an elephant's life the mortality rate is normally around 5 per cent a year; in the crowded herds it is now up to 20 per cent a year. Thus the majority of elephants born fail to reach reproductive age, which begins at around ten to twelve years of age.

Even the surviving adults show signs of disease. Sylvia K. Sikes, a British zoologist, believes that Tsavo elephants are suffering diseases of the heart and arteries as a direct result of the deterioration of their habitat. Her autopsies (on a preliminary cropping of elephants before the Tsavo cropping program ended) comparing the Tsavo elephants with those that had lived in less crowded and more abundant habitats, showed clear differences in the amount of fatty deposits on artery walls. Because in other animals such deposits are chiefly associated with a diet high in animal fats, their appearance in vegetarian elephants suggests other causes. Sikes believes the major causes are prolonged exposure to sunlight due to the destruction of shade trees; overcrowding; a poorer, less varied diet; and frustration of elephant's natural urge to migrate. There is evidence that, under normal conditions, elephants encountering such stresses would migrate to more favorable regions. With the park virtually surrounded by cultivation, the Tsavo elephants have nowhere to go.

Elephants are a problem not only in Tsavo but also in Kabalega Falls (formerly Murchison Falls) National Park in Uganda, Kruger National Park in South Africa, Luangwa Valley Wildlife Refuge in Zambia, and even in Serengeti National Park in Tanzania.

Elephant density alone is not the whole problem. In Tanzania's beautiful Lake Manyara National Park, where Douglas-Hamilton did his research, there are three times as many elephants per square mile as there are in Tsavo. In fact, one theory why the Manyara lions are almost unique in preferring to sleep in trees is that they are trying to keep out of the elephants' way. And yet there appear to be few environmental problems in Manyara. It remains green and thickly vegetated. The difference is rainfall, the single most important factor in almost every African ecosystem. Tsavo can count on between ten and twenty inches of rain in good years, much less in drought years. Manyara gets more than twice as much and is at the base of a major escarpment that gathers water and lets it run into the park.

Virtually the only seasons in East Africa are wet ones and dry ones. In some places, however, the "rainy" season may bring only two or three inches of water. A few miles away it may bring sixty inches. During the rainy season the grass, bushes, and trees are green. But within weeks after the last rainfall, everything dries up. Grass turns brown and many trees drop their leaves. Growth nearly stops, and herbivores must make do with whatever standing hay is left plus the brief blooms of green that follow the rare dry-season shower. In Tsavo even the rainy season's water falls in isolated patches. Biochemical studies have shown that during the dry season, a young elephant's growth virtually stops, much as the growth of trees in the temperate zones stops during winter. Only in the rainy season is the food sufficient to allow growth to resume. Somehow elephants can sense a rainfall miles away. Thus, after each isolated shower, elephants come from miles around to take advantage of the greening while it lasts.

Depending on the amount and distribution of a region's rainfall, it can comfortably support a larger or smaller population of elephants. Tsavo can handle a few thousand—it did for many years—but not the twenty thousand to thirty thousand it now has.

The fact that the Tsavo elephants are trapped is another major factor. Although very little is known about the long-term activities of elephant herds, it is believed that before man became such an intrusive factor, elephants migrated in great herds leaving an area before their destructive habits ruined it and not returning for perhaps decades. Thus there was time for the damaged landscape to regenerate. Today few elephants are able to escape from the parks, the destruction of which continues without respite.

The most widely proposed solution to the parks problem is to cull a proportion of the elephants, reducing their numbers to a level that can coexist with the environment. Elephant-cropping programs have been or are being tried in several places. At Kabalega Falls a few years ago two thousand of some ten thousand elephants were shot. This relieved the pressure at first, but gradually the natural constraints on fertility began to fall. Suddenly relieved of pressure, the elephants began multiplying faster and today have probably already reached their old numbers again. What was at first thought to be a long-term solution was not.

As noted earlier, the prospect of repeated elephant killings in a national park appalls many conservationists, but not in South Africa, however. In Kruger National Park about one thousand elephants are shot every year in a sustained-yield cropping program that appears to be working, not only to keep the elephant herds in proportion to their habitat but also to generate substantial revenue from the sale of ivory and hides. In other parks around Africa too, elephants have been shot to thin out their numbers. But not in Tsavo.

"If it could be done without the objections I see," said

Glover, "I'd crop in Tsavo. But there are just too many problems that I don't think can be overcome. So, we might as well make the best of it and consider this [doing nothing in Tsavo] a control while everybody else tries cropping."

When Laws was heading the Tsavo Research Project he had ideas for making a detailed study of the population structure and breeding patterns of the elephants. Possibly that information and some mathematical computation from the science of population dynamics would show whether the natural decreases in fertility and infant survival would eventually limit elephant numbers. Or, if that seemed too slow a process to save the park, the study could suggest how many and which elephants should be cropped to give the vegetation the best chance to regenerate.

In 1966, along with a former game warden turned elephant cropper, who had made a small fortune selling the ivory and skins of the elephants he shot in Kabalega Falls National Park, Laws shot a sample of three hundred elephants in Tsavo so they could be accurately classified for age and sex and so that their diseases could be diagnosed and pregnancies in progress counted. This, however, was not enough to get the whole story, Laws contended. His belief, based on his studies of elephant distribution in the park, was that there were ten distinct populations that lived in more or less discreet regions of the park. They appeared to have different age structures and therefore could be expected to have different reproductive potentials. Laws said that three hundred elephants would have to be taken from each population to get the full picture. Thus 2,700 more elephants would have to be shot. The outcry killed the plan. Many people doubted the evidence for discrete populations.

"I'm not convinced," said the Kenya Minister of Tourism

and Wildlife, "that it is necessary to kill three thousand elephants in an experiment aimed at seeing how many elephants will eventually have to be killed or whether any will have to be killed at all."

The park warden had, just a few years earlier, conducted a massive military-style antipoaching campaign that put hundreds of people into jail for shooting elephants. He was not prepared to turn around and pay professional hunters to kill three thousand. What would the tourists think seeing elephant corpses all about or even just hearing the gunfire? Of all Africa's animals, elephants are among the most loved by lay conservationists because of their size, intelligence, and close family structure that gives the illusion of adhering to the finest human values.

There were also technical objections on the grounds that the surviving elephants, seeing their fellows slaughtered, would become embittered and turn into man-killers or at least retreat from tourists hoping to photograph them. Laws proposed to meet this objection by shooting entire family groups (a requirement for the research anyway), leaving no witnesses. To no avail. Faced with an untenable situation, Laws resigned his post and left Africa. Nothing further has been done except to gauge in very rough terms the continuing effects of the elephants.

When the severe drought hit Tsavo from April 1970 until late November 1971, most of the park received less than ten inches of rain. Periods of up to five months would go by without a single drop. Elephants began dying by the hundreds, not of thirst for there were still a few permanent water sources, but of starvation. The grasses and leaves they needed were long since gone. In desperation many elephants ate bare twigs and branches, starving to death with their stomachs full of wood. The first to go were the youngest and the oldest. Unable to keep up with the herds as they ranged farther and farther in search of food, the weaker ones died. For most of the little ones that died, their nursing mothers died also. When her calf dies, a mother elephant will rarely abandon it until a long time has passed. Somewhere between two and three thousand nursing mothers died, comprising nearly half the drought's toll.

Month after month, as the drought continued, more and more elephants succumbed. Eventually the toll exceeded the park staff's ability to keep up the count over the vast area. In fact twice as many died as the Laws plan would have taken, and in a manner that yielded very little information of use in preventing further disasters.

In addition to the elephants, Tsavo National Park has the world's largest concentration of rhinoceroses. Perhaps five thousand of this dwindling species lived there until the 1970–71 drought. Some six hundred died then. Had the elephant population been smaller, more rhinos would have lived.

Not long before the drought a private organization had spent hundreds of thousands of dollars to rescue just sixty rhinos from precarious environmental situations in Africa. They were released into Tsavo and other parks. Many of the people who supported the rhino operation opposed the elephant-cropping scheme, but had the same money been spent on Laws's research plan, hundreds of rhinos as well as elephants might have been saved.

Each year since the drought, the normal dry seasons witness more starvations. Before so much vegetation had been destroyed, there was enough to carry the elephants through a normal dry season. But now there is not. Little by little the laws of ecology will see that there are no more elephants surviving than the land can support. In coming years, almost certainly, many thousands more elephants will die in Tsavo. Probably only a few score will go each year until about 1980, when another of the periodic severe droughts is expected. There was one in 1950 when the elephants were still too few to be hurt. There was one in 1960 when some three hundred rhinos died but no elephants. But by 1970 the elephant population was too big, and, as we have seen, thousands died. By 1980 Tsavo could experience an even worse die-off.

It is now probably too late for any elephant-cropping program to save Tsavo's habitat. It is largely gone and would require several decades of almost total protection from elephants to grow back.

"If they wanted to keep Tsavo the way it was," said Walter Leuthold, a Tsavo Research Project staff member, "they should have started shooting elephants ten or fifteen years ago. Now, if they want the old vegetation to come back, I think they'd have to shoot three-quarters of them [at least fifteen thousand elephants]. Three thousand won't do it."

Not everyone sees the loss of the Tsavo trees as a bad thing. Grassland habitats favor a greater variety of large mammals such as antelopes, zebras, and their predators. If Tsavo can have more of these, as well as some elephants, the thinking goes, it will attract more tourists. Already elephants have opened up the habitat so much that plains animals are increasing in numbers.

"Some people think this could be another Serengeti," Glover says. "That's just wishful thinking. The Serengeti has twice as much rain and it's more evenly distributed. We'll get a few more [animals] but Tsavo could never support the herds that people go to Serengeti for."

Still, there is some reason to believe that eventually, perhaps in fifty or a hundred years, the trees may come back to Tsavo. If the elephant numbers diminish and stay low and fires are prevented, seedling trees could grow. It may even have happened once before. In the 1880s, according to the accounts of explorers and archaeological findings, Galla herdsmen lived in the area and there were no elephants to speak of. Since cattle are not kept in forests, Tsavo must have been grassland then, kept open by the grass fires African herdsmen set every year all over the continent to burn off the stubble and encourage new growth. But, for some reason, the herdsmen left the area. Bush and trees must have moved in, followed by a modest number of elephants, creating the Tsavo known in the first half of this century. It may happen again, slowly, but only after man, or nature, does something about the elephants.

Whether Africa's elephants are doomed as a species is debatable. So is the nature of the impact that that event would have on African ecology. Much better understood is the impact of the death of individual elephants on members of their own families. Iain Douglas-Hamilton's observations of this phenomenon, which may have something to do with the notion of an elephant graveyard, have produced some remarkable findings. As with human beings, the death of an elephant has a deep effect on the surviving family members. Douglas-Hamilton recounts an incident in Manyara when he came upon the body of a female who had fallen down a steep slope she was attempting to climb. The cow was lying dead on the slope.

"Next to her stood three calves of different sizes," Douglas-Hamilton writes. "The eldest was moaning quietly but every so often gave vent to a passionate bawl. The second just stood dumbly motionless, its head resting against its mother's body. The smallest calf, less than a year old, made forlorn attempts to suck from her breasts. Then the eldest knelt down and pushed its head and small tusks against the corpse, in a hopeless attempt to move it. I watched them for fifteen minutes repeating these patterns of behavior until suddenly they caught my wind and wandered slowly away."

Harvey Croze, a British biologist who has studied elephants at the Serengeti Research Institute, situated in the park, has made similar observations. Once he saw an old cow dying while her family looked on. When she fell, the younger ones gathered around. Some put their trunks in her mouth, a friendly gesture common within elephant families. Some tried to push her up with their heads or lift her weakened body with their tusks. An independent bull who happened to be with the family at that time took over the efforts to revive her. Failing and perhaps in frustration, he engaged in what may have been an attempt to displace his anxiety. He mounted the dead cow and attempted to copulate. Eventually he moved off with the others. Most of the family had remained with the fallen matriarch for a few hours but one cow stayed till nightfall before leaving.

Park wardens in Africa have reported seeing similar elephant behavior, sometimes noting that the survivors in a family remain with their fallen comrade for days. Elephant mothers have been known to remain with their dead babies for days and there is at least one account of a mother carrying her decomposing infant on her tusks for several days.

Few other animals display such reactions or even any reaction at all to the death of a relative. Zebras, for example, go on calmly grazing within sight of lions killing one of their

herd and feeding on the carcass. Lions display no special reaction to the death of a member of the pride.

The extraordinary response of the elephants, though it
proves nothing for certain, does suggest that these animals
have some sense of mortality. It is well established that elephants recognize one another as individuals and that friendships among particular elephants can endure for many years.
It is also clear that elephants in families receive attention and
help from others; consequently the loss of a family member
may be a genuine disadvantage to the survivors. Efforts to
raise a fallen family member would also have obvious value in
those instances where an elephant has merely collapsed or
fallen without any fatal cause. If an elephant lies the wrong
way on the ground, its great bulk can hinder breathing and
lead to death. The apparently altruistic behavior of others in
helping a fallen elephant to rise could be explained simply as
the standing elephants' attempt to ensure themselves of continuing friendship and assistance. If and when the sick animal recovers, it may well help in rearing the young of other
cows, or, if the sick animal is the matriarch, its decades of
knowledge would obviously be useful in a continuing leadership role.

Yet the story of elephants and death does not end there, for
even when a dead elephant is reduced to bleached bones, living elephants still take unusual notice. Elephants who stumble
on the disarticulated skeleton of an elephant often stop to examine the bones, picking them up, especially the tusks, and
carrying them about. David Sheldrick, the warden of the eastern part of Tsavo National Park, has become convinced over
the years that elephants have a habit of removing the tusks of
dead elephants and carrying them away, sometimes for up to
half a mile. Sometimes they smash the tusks. Although such

accounts have in the past been scoffed at by some elephant authorities, the evidence that this is a regular behavior is now rather convincing. It has even been filmed.

Douglas-Hamilton once conducted an experiment in Manyara by gathering up the remains of a dead elephant—skin, bones, and tusks—and placing them near a river where many elephant families came to drink. He waited in his Land-Rover to see what would happen. "In most cases," he wrote in *Among the Elephants,* "as soon as they became aware of the bones they showed great excitement, raising their tails and half extending their ears, grouping around and carrying out a thorough investigation, picking up some of the bones and turning others with their feet. They usually became huddled in a tight circle so that I couldn't see what they were doing, except when an elephant raised a bone above its head."

In another of Douglas-Hamilton's experiments the living elephants again became excited at the sight of elephant bones. The tusks, particularly, were noticed, picked up, and passed from elephant to elephant. The beasts picked up other bones as well and carried them for some distance before dropping them.

Another curious and possibly related behavior pattern of elephants is the tendency to bury creatures, elephants as well as others, that appear to be dead. This has been observed in the wild and in zoos. Spying an apparent corpse, elephants will gather up sticks and clumps of grass and strew them over the bodies. There is one account of an old women who was in the bush as night fell and who went to sleep under a tree. Later in the night she was awakened by an elephant touching her with its trunk. Terrified, she froze and watched as other elephants arrived and began laying branches over her. When they moved off, the woman was trapped and could not get out from under the heap of branches until a herdsman happened by and heard her calls.

Numerous wildlife researchers have reported seeing elephants bury their own dead or finding dead elephants already covered with branches and dirt. And there is one account of a zoo keeper lying down outside the elephant enclosure to take a lunchtime nap, only to be awakened by clumps of hay hurled over the fence at him by the elephant.

Douglas-Hamilton has considered both the elephant's interest in elephant bones and its burying behavior and concedes that he is unable to explain them. Apparently no one else has come up with a reasonable explanation either. Thus there remains plenty of room for speculation and time for the secret elephant graveyard myth to survive a while longer.

It is very tempting, knowing a little of what elephants are like in the wild, to anthropomorphize their behavior and impute to these huge beasts the rudiments of human thoughts and emotions. There is danger in this, of course, for it can lead to a whole new set of sentimental and unfounded beliefs about elephants.

6

SHARKS

It was the summer of 1975 and the social doldrums prevailed in American beach communities from Montauk to Malibu until, in hundreds of movie theaters across the land, the film *Jaws* opened. Peter Benchley's 1974 novel of the same name, already having sold millions of copies and rapidly becoming an all-time best seller, had been made into a movie that reached additional millions. Americans thrilled, pleasantly or unpleasantly, to the spectacle of a mysterious beast invading the tranquility of a Long Island resort town called Amity and in the space of three weeks attacking unpredictably and killing five people.

Sharks replaced, at least temporarily, all other wild animals as the ones that most engaged people's attention. The image of a huge beast rising vertically from the depths to mangle some unsuspecting swimmer moved from the book's dust jacket and the movie's advertisements to dozens of political cartoons. Inflation became a killer shark about to destroy a swimming John Q. Public. Communism became a ravenous beast about to carry off Portugal. A shark labeled Ronald Reagan became a rising threat to President Gerald Ford in the race for the 1976 Republican presidential nomination. Entrepreneurs advertised real sharks' teeth in national magazines. Shark T-shirts became a national fad. Weekend sailors painted the prows of their skiffs and runabouts with gaping jaws and gleaming triangular teeth. Imitation shark dorsal fins appeared in swimming pools and reflecting ponds a thousand miles from any ocean. Some bathers even stayed on the beach or, if they ventured into the water, thought themselves quite daring.

Jawsmania, as *Time* magazine called it, is worth remarking in the context of this book for two reasons.

The first is that, all the dangerous large animals being pretty well wiped out from populated parts of the United States, the vast majority of Americans have never had to contemplate the thought of living in proximity to a man-eating animal. The wolves that colonial villages two centuries ago imagined to be a threat are gone from the urban East. The bears, though potentially dangerous, have been reduced in most people's minds to national park clowns. The mountain lions, never much as man-eaters, are practically exterminated. As far as many Americans were concerned, the wilderness had been tamed. Then, all of a sudden, they were reminded that just yards off those sun-washed beaches might lurk a deadly creature still waiting from some prehistoric era to kill them.

The second reason Jawsmania is worth remarking is that nobody really knows much about sharks. The unknown is, of course, at the root of most fears. Despite decades of good work by marine biologists, the U. S. Navy, and many others, sharks remain among the least understood of any large animal. Responsible writings on sharks have run the gamut from declarations that this beast exists only to kill and kill in the most grisly fashion to pronouncements that sharks are nothing more than dull, stupid scavengers whose threat to man is grossly exaggerated.

For example, the American naturalist Roger Caras, in his book *Dangerous to Man* (1964), contends the shark is the world's deadliest man-eater. "The shark," he also says, "is a beautifully-designed predator. It is a killing machine with a voracious appetite." Caras, who is generally known as a champion of humane treatment of animals, recounts with obvious glee some occasions in which he had the opportunity personally to destroy a number of sharks.

On the other hand, there is the view of the late William Beebe, the American ornithologist and underwater adventurer, who once wrote: "The ability of the human imagination to see what it thinks it ought to see is astonishing. As long as my book- and legend-induced fear of sharks dominated, I saw them as sinuous, crafty, sinister, cruel-mouthed, sneering. When I came at last to know them for harmless scavengers, all these characteristics slipped away, and I saw them as they really are—indolent, awkward, chinless cowards."

What, then, is the truth about sharks? Here one would want

reports from observers of free-living sharks because they, of course, are the only ones that attack swimmers. In the oceans, it is often reported, sharks are the prime enemy, aside from man, of dolphins, or porpoises. Yet in the many giant aquariums where the two are kept together and under almost continual observation, shark attacks on dolphins almost never occur. Either the reports from ocean-going observers are wrong or sharks don't behave in captivity the way they do in the wild. The latter is quite possibly the case. Unfortunately, there aren't any solid behavioral studies of free-living sharks. Nearly all the information that follows is based on scattered accidental observations and a few efforts to collect such observations and evaluate them as a body.

There are more than three hundred known species of shark and undoubtedly a number of unknown species. As adults they range in size from less than a foot in length to more than fifty feet. Among the smaller sharks is the spiny dogfish, for some Britishers the first half of the traditional fish and chips. The two biggest sharks, the basking shark and the whale shark, are not known to attack any sizable creatures. Rather, they feed on plankton, the drifting minute plant and animal life of the oceans. Both these beasts are known as rather torpid creatures, and sportsmen's books contain many accounts of swimmers getting in the water with, say, a whale shark and hitching a ride on its fins or even calmly peering inside its gaping jaws.

The villain in *Jaws* was a smaller shark than these—the white shark, or, as storytellers prefer, the *great* white shark. Most scientific and sporting accounts fairly well agree that this is indeed the most feared and fearsome of fish. That it is large is amply documented by measurements of captured specimens. White sharks over twenty feet in length—the size of the creature in *Jaws*—have been caught and estimates based on sightings at sea have ranged up to forty feet. A twenty-one-footer taken off Cuba in the 1940s weighed 7,302 pounds. That these animals can be man-eaters is well established. Consider, for example, what happened in July 1916 at the New Jersey shore.

A twenty-four-year-old man was swimming in five feet of water when an 8½-foot-long white shark approached and clamped its jaws around one leg. Other swimmers came to the man's rescue and pulled him toward the beach with the shark still hanging on. Not until they were almost ashore, in eighteen inches of water, did the fish leg go and swim back to deeper water. In a hospital, a few hours later, the man died. Four days later, not far away, another man was attacked by a white shark, also estimated at 8½ feet. It ripped off both the man's feet and stripped the flesh from much of what remained of the man's left leg. He died on the beach. Then, six days later the white shark struck again, killing a boy and a man who attempted to rescue the boy. Later the same day, about half a mile away, the shark ripped off another boy's leg. The child survived. The basic premise for Peter Benchley's novel, then, is factual.

In 1960, again off the New Jersey shore, sharks attacked three times in a nine-day period in August. All three victims survived. The shark or sharks were not identified.

Many other attacks on people by white sharks have been recorded. There are also numerous accounts of white sharks charging into boats and even biting them, leaving fragments of or whole teeth embedded in the hulls.

There is even a longstanding belief that the great fish that swallowed Jonah was not a whale, as stated in the Gospel of St. Matthew, but a white shark. No less a figure than Linnaeus, the great eighteenth-century Swedish botanist who devised the present scientific system of naming living species, appended a note to his description of the white shark in *Systema Naturae* (1735) that said, "It is likely that the prophet Jonah remained in the belly of this animal for a space of three days, as Hercules of old did for three nights." Sharks commonly gulp down their prey whole or in big pieces, and the stomachs of dead sharks have often been found to include recognizable fragments of human bodies, such as whole limbs. But prophets or other human beings most assuredly are not the main item in the diet of a white shark or any other shark.

The fact that sharks occasionally eat a human being probably reflects not so much a preference for human flesh as an indiscriminate taste in food: sharks will eat practically anything. In the stomachs of dead sharks people have also found rubber boots, life preservers, dogs, coconuts, rudders, and even some fish. Beyond their eclectic dietary habits, however, practically no generalization about shark behavior has stood up.

The ichthyologist Leonard P. Schultz, of the Smithsonian Institution and long the head of the Shark Research Panel of the American Institute of Biological Sciences, once analyzed reports of 1,406 shark attacks dating from the 1850s. He found that the beasts struck on sunny days and cloudy days; in clear, still water and in muddy, roiling water; at night and during the day; in lakes, bays, and rivers as well as the open ocean; and that the attacks occurred at all seasons.

As Matt Hooper, the marine biologist in *Jaws* who tries to help Amity, remarks in the book, "Scientists spend their lives trying to find answers about sharks, and as soon as they come up with a nice pat generalization, something shoots it down."

Like much of what the book *Jaws* says about sharks, that is true. But the impression many people have taken away from the *Jaws* story—that sharks are a significant threat at the beach—is definitely not true. One possible exception is Australia where special shark surveillance measures are used at some beaches. In fact, compared with practically any other danger to human life or limb, sharks are among the least of things to worry about—even for ocean bathers. Many times more people are killed by the undertow than by sharks.

The Shark Research Panel, which has kept track of reported shark attacks since its formation in 1958, gets an average of about fifty such reports each year from all over the world. In one five-year period, for example, there were 161 shark attacks on ocean bathers near the shore. Of the 161 attacks, thirty-two occurred in United States waters, an average of less than seven a year. Few were fatal, but even if all were, this would be less than the number of deaths in the United States due to aspirin overdose or lightning strikes.

Try to make a gripping adventure story about a great white aspirin tablet or a rampaging thunderstorm. The notion is ridiculous not because sharks are genuinely a bigger threat but because as wild animals that still frequent American waters, they are practically the only sizable creature left that can still elicit in Americans that primal dread of the wild beast—an emotion that remains common only in parts of Africa or a few places in Asia or South America.

In response to that primal emotion elicited during the wave of Jawsmania, some Americans responded in classical fashion —they wanted to kill the dangerous beast. Shark fishing suddenly grew in popularity, even among people who had not regarded themselves as fishermen or fisherwomen before. Hundreds, perhaps thousands, of sharks of all kinds were hooked or gaffed. Some were slit open at sea and tossed back to provide the bloody spectacle of the blood-crazed shark eating itself. Many were brought back to shore as proofs of manhood—the brave hunter subduing the vicious beast for the good of society. Quite often the jaws were cut out as trophies while the lifeless bodies were either left to rot or thrown back into the water.

In the summer of 1975 Jawsmania was interpreted in the press as an amusing national fad, a mindless eccentricity spawned to enliven a time of national ennui. In fact, it was the resurgence of a basic, atavistic emotion—dread of the mysterious wild beast. It was probably the last such opportunity for most Americans in their own land.

7

BABOONS

Baboon. The word even sounds vulgar, grotesque. The image is of a scrappy, vile, brutish little monkey. Most Westerners see them only in zoos where they seem to do nothing but pick through each other's hair and masturbate.

Like hyenas, baboons in Africa have until recently been classed as vermin and been the targets of innumerable extermination campaigns. A favorite pastime of some national park wardens, especially those trying to grow vegetables in their compounds, has been to sit on the verandas of their

houses and shoot any baboons that appear. Poisoning, however, is more common and many thousands of baboons have
sickened and died in the bush after feasting on the tainted
offerings of a national park warden.

Baboon poisoning has gone out of fashion in recent years,
because of objections from conservationists and tourists but it
used to be one of the official duties of vermin control officers
to wipe out as many as they could. Captain A. Ritchie, head
of the Kenya Game Department in the 1920s and 1930s once
described one of his more outstanding vermin control officers
by saying, in an annual report, "He worked well and consistently, poisoning hyena, baboon, bushpig, leopard and por-
cupine."

Robert Ardrey, an American amateur naturalist and author, whose books have done much to direct Western attention to Africa, its animals, and evidence of early man's evolution and to promote the notion that man's presumed violent
streak derives from his African genesis, doesn't think much of
baboons either:

"He is a born bully, a born criminal, a born candidate for
the hangman's noose," Ardrey wrote in *The Territorial Im-
perative* (1966), revealing his dominant personal belief that
vicious behavior is innate in man and beast. "As compared
with the gorilla, the baboon represents Nature's most lasting
challenge to the police state. He is as submissive as a truck, as
inoffensive as a bulldozer, as gentle as a power-driven lawn-
mower. He is ugly. He has the yellow-to-amber eyes that one
associates with the riverboat gambler. He has predatory in-
clinations, and in certain seasons he enjoys nothing better than
killing and devouring the newborn fawns of the delicate gazelle. And he will steal anything."

Ardrey, presumably, would not have disagreed with the

vermin control officers' doing their duty. Herein lies one of the ironies of current thinking about wildlife. For Ardrey based his understanding of baboons at least partly ón several thoroughly scientific and intensive studies of free-living baboons—among them the published findings of Sherwood Washburn and Irven De Vore, two first-rate American anthropologists who, as a team, studied baboons for ten months in 1959 and 1960 in Nairobi National Park, in Kenya. Their study of baboons has long been considered definitive.

Ardrey relied on the Washburn-De Vore study and wound up disliking baboons as much as did those who poisoned them for no scientific reasons at all or who only saw them in zoos. At first this seems like a case of scientific study confirming popular mythology. And, as far as De Vore and Washburn went, that seemed to be so. They reported that baboon troops were organized into a tight structure dominated by a tyrannical coalition of two or more big males. Ardrey called them a "gang of implacable thugs at the top." Below this there were individual rank orders. De Vore and Washburn concluded that the "central hierarchy" ran practically everything that took place within the baboon troop. They directed its travels, stopped fighting among lower ranking animals, and protected the troop from predators. They said the central hierarchy had this omnipotence because it remained stable for long periods of time. They also suggested that a baboon's life was one long scramble for status within the troop. Baboons, especially the males, were pictured as almost constantly clashing in power struggles, like hoodlums vying for control of a street gang. The study was fine except for one thing. It turned out not to be true of baboons everywhere.

In the years since Washburn and De Vore made their study, many other researchers have observed baboon troops in other places and found very different social structures. It is now recognized that baboon behavior is not immutable. Baboons are not "born" anythings. Depending on the habitat—and baboons occupy very diverse environments, from open grassland to densely forested mountains—and on the size of the troop, the behaviors and social structures may be as different as anything seen in human societies.

This variation of a species from population to population probably is true, at least to a small extent, of every animal species discussed in this book, and this should be kept in mind before one assumes that all members of any species necessarily behave exactly alike. This fact is sometimes forgotten by budding wildlife biologists who are trying to select a species to concentrate on. It is often heard that "lions have been done" or "What else is there to do on chimpanzees that Jane Goodall hasn't done?" There is every good reason to study the same species in different settings for any differences in behavior that may be found may tell us as much about the species as did the original findings. Such differences can mean that a species is far more intelligent or adaptable than had been supposed. It may be that baboons, for example, have more room in their brains for variable behavior than the nonprimates. For example, De Vore and Washburn counted thirty-two body gestures (yawns, grins, lip smackings, etc.) and fourteen vocalizations used for communication among the Nairobi baboons. This appears to be a richer communications ability than most lower animals have.

Putting all baboon studies together, it seems reasonable to conclude that different populations of baboons, like elephants, have what might be called "differing cultural traditions" that make one population act like a street gang, but another like a church choir.

Another variable tradition is diet and how it is gotten. De Vore and Washburn's baboon troop were overwhelmingly vegetarians but did eat insects, eggs, and an occasional mammal about 2 per cent of the time. However, long-term observations of baboons living freely on a forty-five-thousand-acre farm named Kekopey, near Gilgil, Kenya, indicates those baboons eat meat far more often. In fact, over a period of years the rate of killing prey for meat at Kekopey appears to have been growing, as if a new tradition is spreading or becoming more intensively practiced. The baboons kill birds, rabbits, gazelles, and the young of other antelopes up to the size of impalas, of which there are thousands at Kekopey.

In 1970 and 1971, when Robert S. O. Harding, an anthropologist at the University of Pennsylvania, was studying the Kekopey troop, he counted forty-seven kills in 1,032 hours of observation; adult males did all the killing and nearly half the eating. By 1973, when Shirley C. Strum, an anthropologist at the University of California, San Diego, had taken over observations on the same troop, the predatory tradition had grown. She saw one hundred kills in 1,200 hours of observation; both males and females alike were doing the killing and baboons of all ages and both sexes were eating. By 1974 the pattern of killing had also changed. In the earlier years, the baboons killed mostly animals they had come across inadvertently, such as sleeping fawns. Now they were going on deliberate hunts, sometimes spending two hours at a stretch searching for herds of gazelles, running at them to "test" the herd just as wolves do, and grabbing the slow or weak ones.

The researchers also witnessed the accidental discovery by baboons of a new hunting strategy and its adoption as a regular method. Initially individual baboons hunted and killed their prey and ate alone. Then it developed that once a kill had been made, other baboons would converge on the scene of the kill for a share of the meat. One day a lone male was chasing a gazelle. Three other males noticed the action and were walking toward the scene to watch. When the original baboon was about to give up the chase as fruitless, the gazelle inadvertently ran toward the three approaching spectators. Seizing the opportunity, one of the spectators raced after the gazelle, chased it a distance, and then appeared to tire. Just as it did, the second of the spectators took up the chase. The gazelle easily outdistanced this baboon but suddenly changed direction and accidentally ran into the third spectator. The startled baboon hesitated a moment and then bit the gazelle's underbelly.

"From that point on," Harding and Strum reported, "the male baboons gradually adopted this relay system as a regular stratagem, chasing their prey toward a nearby male instead of out on the open plain. Such joint ventures appeared to be more successful than those carried out by lone males."

The researchers also found that in the early days of the hunting tradition, baboons could not distinguish between all-male herds of gazelles and herds that included females. The bachelor herds would have none of the younger, easier-to-catch gazelles such as those who stayed with their mothers, and efforts to make kills among the all-male herds failed. Later, however, the baboons learned to tell the difference between gazelle herds and ignored bachelor herds.

That antelope killing and eating are not inevitable baboon behaviors seems clear from comparisons of the Kekopey study troop with those made elsewhere. Other troops have been known to catch and eat rock hyraxes, small mammals which look vaguely like guinea pigs, and to ignore gazelles. There are hyraxes in Kekopey, but the baboons studied by Harding and Strum were never seen to hunt or catch them. Another troop at Kekopey is known to catch and eat guinea fowl, but the study troop let these birds walk through their midst without paying the slightest attention. It seems inescapable that these dietary preferences and hunting methods are purely cultural phenomena that call into question any assertion that killing behavior in higher animals is necessarily innate.

The Kekopey studies also found no "central hierarchy," although the methods of study, which differed from those of De Vore and Washburn, may have masked this. Instead of keeping close track of which animal fought with which and who won, Harding and Strum concentrated on charting "friendships" among baboons—which individuals associated with one another habitually. Thus, while Washburn and De Vore concluded that a strange baboon trying to enter a troop had to fight his way up the dominance ladder to become accepted, the Kekopey studies suggest that strangers join their troop by becoming friends with individuals on the periphery of the troop's social structure and using these friendships to gain others. This difference may reflect cultural variations between the two troops or variances in the philosophical approaches of the researchers.

Another way in which cultural traditions vary among baboons involves how they behave when encountering threatening enemies from another species. There is one report of baboons picking up rocks and tossing them over a hillside in the general direction of human beings trying to approach them. The use of projectiles by baboons is otherwise unknown; more typically baboons try to run away. But even the running away may take different forms. Threatened baboons usually climb trees to escape. But in areas where people have shot at baboons for many years, the baboons run into the bush for cover instead of climbing trees. Those already in trees, on seeing people, will climb down and try to slip away into the bush. The same animals, threatened by a lion or other carnivore, will go up the tree to escape. Apparently baboons have learned, or perhaps their ancestors learned and passed on the information, that people can shoot them in trees.

Another deviation from De Vore and Washburn's findings in the Nairobi park involves the behavior of the adult males when a nonhuman predator approaches. In Nairobi the big males put themselves between their troop and the danger, following the troop as it moved to safety and frequently glancing back. A study in Uganda, on the other hand, showed that the males were the first to run.

There are many other things that could be said of baboons. In fact, it would not be difficult to comb through the various studies in print and select facts and anecdotes that, arranged in a particular way, would present baboons as models of almost any kind of society imaginable. You could find baboon troops that were "dictatorships" and others that were "democratic." There are baboon troops where the big, powerful members seem to take special care to share food with the smaller, weaker members; but there are others in which the powerful stay that way by trampling on the rights and needs of the weak. In other words, one sees almost the same diversity that exists among human societies.

The comparison of any animal with human beings is always risky, as much of this book is trying to show. But there is a little more reason for attempting the comparison with baboons than with most other species. Baboons and people are both primates, members of a very broad group that includes apes, monkeys, lemurs along with other species. Baboons, being monkeys, are higher primates. Among all the primates, there are only two kinds that have taken to life on the savanna, the tree-dotted, predator-infested grassland that comprises so much of East Africa. They are baboons and human beings. The chimpanzees, our closest anatomical relatives, stay in the forest. Baboons and people are also omnivorous: both species eat meat and vegetables. Both live in fairly close-knit family groups. Although baboons walk on all fours, their hands are dextrous enough to pluck grains of sand out of their eyes. And both baboons and human beings have an ability to adapt their ways of life to various environments and to adopt better ways of doing something when they are found. It is not surprising, then, that many anthropologists have suggested that baboons might be able to teach us something about the life of our early ancestors. They may well be more apt models of early man than the apes.

Is there something about the life styles—the vulnerable grassland life or perhaps about a mixed diet that calls for some individuals to be hunters and others to be gatherers—that leads intelligent creatures to form the kinds of well-regulated societies that baboons and human beings do? Is there something about these factors, combined with a broad repertoire of adaptable behavior patterns, that leads people and baboons, depending on the circumstances, to form the diverse types of societies that they have?

Such questions may be far-fetched, but could it be that it is at least our dim perception of the possible similarities that makes us dislike baboons so? If we knew these intelligent, versatile beasts better, however, might we not also gain a little more appreciation of ourselves?

8

HYENAS

Walt Disney never made an animal film about cuddly hyenas. F. A. O. Schwarz doesn't sell cute little stuffed hyenas. The World Wildlife Fund has never issued an appeal for donations to rescue or protect threatened hyenas. Tourists who spend thousands of dollars to come to East Africa to see animals are quite happy to go home without ever having seen a hyena. The hyena doesn't make an appearance in any Western children's story, not even in the role of villain. In Africa, one of the two areas where the beast lives (the other is India), it does figure in some traditional stories, but usually as an object of ridicule or as an embodiment of evil.

Hyenas, to put it mildly, are not box office. In fact, until quite recently, even national park wardens in Africa, particularly the British wardens, boasted of the service they had performed by shooting or poisoning hyenas, exterminating them as dutifully as a housekeeper spraying the kitchen for cockroaches. Almost nobody loves a hyena.

This is not to say that now and again you won't hear some ecologically minded person point out that the natural world needs scavengers, creatures whose place in nature is to go around eating up the rotting carcasses that, it is claimed, might otherwise spoil the wilderness. Nature's garbage men, they are called in some places.

Consult almost any reference work or any other printed source and you will be told two things about hyenas. They are scavengers and they are cowardly. Even the generally authoritative. Encyclopaedia Britannica, source book for countless high school term papers, says, "Hyenas are principally scavengers and in many regions depend largely on the remains of carcasses left by lions. Although they are generally quite cowardly, they will attack helpless animals and may become quite daring when hungry."

However misshapen, scruffy, cowardly, smelly, and loathsome hyenas are said to be, some naturalists have argued that they are useful. Where there are predators, you need scavengers. That is quite true, but people have mistakenly assumed that those terms—"predator" and "scavenger"—define an animal species. Actually, they define a kind of behavior. The lion is a predator sometimes, but at other times it is a scavenger. The hyena, to be sure, scavenges sometimes, but it can also be a predator. In fact, in areas where both species exist, it is the hyena which is predominantly the predator and the lion which is more often the scavenger. In many parts of East Africa— the Serengeti, in Tanzania, famous for its lions, and nearby Ngorongoro Crater—hyenas are the most numerous predators, and it is they who do most of the killing, with the lions doing most of the scavenging. In Ngorongoro hyenas do virtually *all* (93 per cent) of the hunting and killing and lions do nearly all the scavenging.

And, what's more, the prey that hyenas select are not usually small or "helpless" creatures. They take full-grown wildebeests, sleek and healthy zebras, and, now and then, African buffaloes weighing half a ton. They do not do this by accidentally stumbling upon such a creature while it is sleeping. In hunting wildebeest for example, the "cowardly" hyena, working alone, approaches a grazing herd, dashes into its midst briefly in order to scatter the animals, selects an individual, and gives chase. Racing along at speeds of up to forty miles per hour—almost as fast a thoroughbred horse—and sometimes continuing the chase for three miles, the hyena closes in on its tiring target. When the prey is within range, the hyena's powerful jaws clamp into the animal's flesh, ripping away with such ferocity that in less than a minute the wildebeest is literally in a state of paralyzing shock. Sprawled immobile on the ground, the animal is quickly killed by the hyena, often by disembowelment, and eaten.

If that is a typical scenario, and it is for wildebeest (hyenas use different methods for other prey species), where did the idea come from that hyenas do not hunt their own meat on the hoof? The answer can only be embarrassing for the predecessors of today's wildlife biologists who thought they knew how hyenas lived. Hyenas normally hunt at night when most people are asleep. By daybreak, when the average field biologist or tourist or big game hunter has rolled out of bed and set off in search of game, it is likely that a pride of hungry lions, attracted to the kill by the typical whooping and cackling (the famous laughing) of hyenas, have driven off the partly-stuffed hyenas and stolen the kill to scavenge the remains. By the time the people come along, the lions are feeding and two or three hyenas, still hungry, are circling warily, waiting for another chance at the carcass. A few vultures and jackals may have shown up to wait their turns. To someone happening on the scene at this moment, the most obvious interpretation, given the prevalent beliefs about the hyena, could only serve to reinforce the myth of the hyena's "loathsome" carrion-eating habits.

Of course, the definition of "carrion" has never been very clear. The decay processes begin within hours of an animal's death. One effect of this breakdown of tissues, as any butcher who ages meat for his more discerning customers knows, is to tenderize the meat. If one is going to think less of a species for carrion eating, then one must dethrone not only the lion but also the leopard, that sleek and sinewy cat of the night that is such a favorite of lay conservationists. In fact, unlike most other predators, leopards often prefer their meat ripe and will store a kill high in a tree for a few days before eating it. Hunters know this and can easily shoot a leopard after putting out a bait of rotten meat.

It was not until the early 1970s that the negative view of the hyena began to change, through the work of one man, the young Dutch animal-behavior researcher Hans Kruuk, who went to Africa and spent three and a half years in the Ngorongoro Crater systematically observing hyena behavior. Beyond his scientific acumen, developed like that of elephant researcher Iain Douglas-Hamilton under Nikolaas Tinbergen,

Kruuk brought one crucial asset to his studies. He was willing to stay up nights and see what hyenas do then. He also kept a pet hyena. The result of his studies is an extraordinary book, *The Spotted Hyena* (1972), which, although a detailed scientific monograph, makes fascinating reading. It is, along with a handful of scientific papers by Kruuk, virtually the only source of documented truth about hyenas. In addition to the hyena's predatory behavior, Kruuk also established that hyenas, widely regarded as mostly solitary animals, are in fact highly gregarious and live much of the time in clans of up to eighty individuals, with clear-cut social structures and rituals. He also showed that unlike most other animals the females are larger than and dominant over the males.

After Kruuk completed his work in Ngorongoro, Jane Goodall, the young British naturalist better known for her work with chimpanzees, carried out further studies on one hyena clan, confirming for the most part Kruuk's findings. A highly personalized version of her observations appears as a chapter in the book she coauthored with her husband, Hugo van Lawick, *Innocent Killers* (1973).

The hyena does not look like most people's image of a lovable or a handsome animal. Its head and jaws are massive. The front legs and shoulders are big and powerful but the animal's back slopes down to a disproportionately small set of hindquarters. The coat is scruffy, almost moth-eaten in appearance, and around the neck and shoulders it sticks out in unsightly tufts. By far the hyena's oddest anatomical feature is that the genitalia of the male and female cannot be readily distinguished on sight. The female's clitoris is enlarged to equal the size and shape of the male's penis and is capable of erection to the same degree. Behind the clitoris are two sacs filled with fibrous tissue that look just like the male's scrotum.

This extraordinary situation long ago led to the belief that hyenas were hermaphrodites, each individual possessing the sexual apparatus of both sexes. Nobody bothered to dissect the hyenas that didn't give birth to see whether they lacked uteruses. This mistaken belief goes back more than 2,300 years to Aristotle. Solid scientific evidence on the matter was not put forth until 1939, but the myth still crops up in more recent nonscientific literature.

Kruuk says that the best way he could tell the males from the females when observing hyenas in the wild was to look for the female's larger nipples. In some cases he could distinguish a slight difference of appearance between a real scrotum and a "sham" scrotum.

The reason for this anatomical peculiarity is not entirely clear, but the genitals do play an important role in the meeting ritual of hyenas that have been separated for a time or between two that are strangers. As the animals approach one another, one of them typically shows signs of fear. Usually it is the smaller that does this—a cub meeting an adult or a male meeting a female. The subdominant animal approaches cautiously, often with its tail between its legs, its ears low, and posture betraying fear. As the two meet, they sniff each other's head for a few seconds. The penis or clitoris of each animal becomes erect. Sometimes that is all they do and both animals appear to be reassured—they are still friends and on good terms—and they go on about their business. But, at other times, and particularly when the two hyenas have been separated for a long time, the meeting ritual goes another step. The animals move parallel to one another, facing in opposite directions and the lower-ranking animal lifts one hind leg, allowing the other to sniff its genitals. Then the dominant animal does the same. This mutual sniffing, and sometimes licking of genitals, goes on for ten or fifteen seconds and then the two part company.

Kruuk, who discovered all this, speculates that the prime role of the female's enlarged clitoris is in this ceremony. He established that it does not become erect during mating. No one knows whether female dominance of the hyena clan was a cause or an effect of the penislike clitoris, but Kruuk speculates that it evolved because of the hyena's way of life—living primarily in a close-knit group but occasionally leaving the clan for days at a time to roam. Creatures that cannot make promises or sign contracts must have some form of ritual whereby, after each parting, it is possible to be reunited on intimate terms. The genitals are obviously quite vulnerable to attack during the ceremony and exposing them to the teeth of another hyena rather clearly indicates a willingness to trust and a desire to be trusted. It is significant in this interpretation that it is always the lower-ranking animal that lifts its leg first. The ritual may also be a way of patching up quarrels, for it often takes place immediately after one hyena has snapped at another for some reason.

The structure of hyena society varies from place to place, but its major unit is the clan, a more or less permanent coalition of, typically, from forty to sixty hyenas that live within a clearly demarcated territory of ten or fifteen square miles. Many members of the clan are related, but the group commonly includes immigrants from other clans. Within the clan there may be several smaller groups, each with its own den— a system of underground tunnels where the cubs are sheltered and where the adults may go to sleep during the heat of the day.

Although hyenas sometimes do things as solitary animals, ranging for many miles and many days away from the den from time to time, they are principally social animals. In fact, Kruuk theorized that some of the standard hyena behaviors he saw were performed "with the sole function of 'doing something together.'" An example was what Kruuk called "social sniffing."

"Typically," he wrote in *The Spotted Hyena*, "a group of hyenas would walk together at a brisk pace, fairly close, several with their tails up. Suddenly one would stop, sniffing the ground at a place apparently randomly chosen—immediately all the others came rushing up, all starting to sniff the same place or immediately next to it. While sniffing they appeared very excited, with their tails up and making many fast head movements. One important aspect of the picture was that the participants touch each other while sniffing, mostly with their necks; often one hyena would lean over the neck of another to sniff the ground or lean sideways against the other."

Then, after half a minute or so, the animals would resume their travels, only to repeat the scene every five or ten minutes. Kruuk could determine nothing special about the places where the social sniffing took place. He found he could even induce his pet hyena, Solomon, to do some social sniffing by moving his hand over a randomly chosen place on the ground in excited movements imitative of hyenas. Solomon readily picked up the message and started happily sniffing the ground and rubbing against Kruuk's arm as he did.

Although the behavior has the look of something done for the sheer joy of being out and about with the rest of "the fellas," Kruuk notes that it probably serves a more serious purpose as well—keeping the group friendly and well mannered. Hyenas are powerful carnivores and sometimes fight one another to the death. Furthermore, since the success of a hyena clan depends on working well as a group, anything that promotes group cohesion and amity has survival value for the group.

Although these rituals keep peace within the clan, relations between neighboring clans are not so pleasant and many clan activities of hyenas are calculated to minimize contact with outsiders. For example, if a prey animal being pursued by one clan of hyenas happens to cross the boundary into another clan's territory, the hyenas, like county sheriffs, will often quit the chase at the border. If they do enter the foreign land and make the kill, the commotion is likely to attract the resident hyenas. Their appearance is frequently enough to send the invaders scurrying home. Kruuk cites numerous instances of hyenas pulling down and killing an animal in alien territory, only to yield the uneaten carcass to the resident hyenas. If the kill is made only a short way over the border and there are more invading hyenas than residents, a loud and raucous dispute may ensue. As with most territorial animals, it makes a difference how far into alien regions one penetrates. Near the border an invader acts more boldly, but deep within enemy territory, even a goodly number of invaders can be put to flight by a few residents.

Kruuk observed numerous wranglings between neighboring clans. "Generally," he says, "a clash consists of a great deal of calling and displaying and chasing, and physical contact is rarely made. But if it is, members of either side may be severely mauled or even killed."

Out of 109 kills of prey that Kruuk witnessed, thirty-six led to a fight with neighboring clans. On four such occasions a hyena was killed. Once a hyena is dead, other hyenas will eat it, not within minutes, as they might consume a wildebeest or a zebra, but gradually over the course of a day or so.

The boundary between clan territories is, to a hyena, as

clear and unmistakable as any national boundary observed by human beings. The line is marked by various forms of scent deposits, which are frequently strong enough for human beings also to discern.

The establishment and maintenance of the boundary is, along with zebra hunting, one of the clearest proofs that hyenas, contrary to popular supposition, are highly intelligent and well-organized social animals. Indeed, their activities are typically so well organized that when a group sets out to conduct a routine border patrol, it totally ignores convenient prey animals. In the same way, when a group of hyenas organizes to conduct a zebra hunt, in which special tactics are used, the predators will ignore prey of any other species.

Let us look first at the border patrol. A group of eight or ten hyenas from a clan will assemble at a regular spot, which Kruuk called "the club," and indulge in a little social sniffing and other acts of group camaraderie. Then, as if on signal, the dominant female trots off, leading the group directly to the edge of its territory. The boundary, as just explained, is recognizable by previously deposited scents, but to renew the markings, the hyenas make use of two types of scent glands. One set is situated just inside the anus and can be exposed for use by everting the rectum slightly. The hyena selects a tuft of tall grass, straddles it and walks forward slowly, pressing the stalks between its hind legs and the everted rectum. The gland's thick, whitish secretion is "pasted" onto the stalk, giving it, in Kruuk's estimate, "the smell of cheap soap boiling or burning." The other set of scent glands is in the paws, between the toes. To deposit this material, hyenas scratch furiously at the ground.

After the border patrol has pasted and pawed at one spot on the border, it moves down the line, paw glands depositing more scent as the group walks, and stops at another spot for more concentrated pasting and pawing. Even if the hyenas come upon convenient prey animals, they pay them no attention and keep on with the patrol, typically covering a mile or so of boundary before quitting for the day.

Although only a freshly marked boundary can be smelled by human beings, old scent markings are readily apparent to hyenas because their sense of smell is very keen. In fact, despite the hyena's reputation as a creature with no admirable talents, its senses of hearing and vision also are most acute. Kruuk estimated that in the daytime he could see about as well as hyenas could but that at night hyenas are far superior. "For instance," he notes, "they could, at night, recognize another hyena [determining whether he belonged to the same clan or not] even when the other hyena was coming from downwind in complete silence, whereas I could hardly see that there was an animal at all. Hyenas were able to react immediately to a small disturbance in a herd of wildebeest miles away when we could hardly see the herd with the naked eye." As to their hearing ability, Kruuk reports that "a hyena lying down at night might suddenly jump up and run in a particular direction (not necessarily upwind) toward a place where a group of hyenas were engaged in the noisy consumption of a kill—sometimes we ourselves could definitely not hear these sounds from the place where the hyena was lying."

But, of all the hyena's behaviors, nothing more surely flies in the face of centuries of myth about the hyena than its zebra hunting technique. It is a superb example not only of the hyena's sophisticated social structure but also of its masterful hunting ability. Zebras are, after all, wild horses, with all the strength, speed, and intelligence of horses anywhere. And zebras have a tight social grouping in which the herd is defended by a dominant stallion that is not afraid to attack any predator, be it hyena or lion. The hyena's zebra hunt, like its border patrol, is an example of the hyena's ability to set a goal and plan ahead to accomplish it. Unlike the killing of wildebeest, which is usually done by one, sometimes two, hyenas, killing a zebra requires the hyenas to form a co-ordinated pack. After numerous observations, almost invariably at night, Kruuk concluded that hyenas had to decide they wanted zebra well in advance of attacking the prey. Kruuk became so familiar with the hyenas that he could tell, when he saw a group of hyenas walking, whether they were out for zebra.

The zebra hunt begins much like the border patrol, with a group of hyenas, averaging ten or eleven individuals, gathering after sundown at "the club" for social sniffing and a bit of pasting and pawing. Then, under the leadership of the largest female, the group moves off. Even if wildebeest or gazelles or any other prey species is nearby, the hyenas ignore them, sometimes actually walking through a herd of wildebeest to find what they want. When they finally see zebras, the hyenas move close together and approach slowly, looking almost as if they had no interest in the zebras.

"When the hyenas were very close," Kruuk recounts from his field notes on a typical hunt, "the zebra stopped grazing and closed up together, their heads up. When the hyenas were about four meters away, the stallion turned toward them and charged, head low and teeth bared. The hyenas scattered out of his way and the stallion immediately turned back to his family. The zebra bunched up and began running slowly (at a speed of 20 to 25 kmph) away from the hyenas. The stallion ran just behind his family; several times he charged the hyenas and tried to bite them, and once he kicked out with his hind legs. The eight hyenas galloped just behind the zebra, five of them close around the stallion and the other three just in front of him. Several times some of the zebra barked excitedly.

"When the stallion was a little farther behind than usual, the rest of the family made a 90-degree turn, bunched up till they were almost touching each other, and virtually stopped; again the stallion attacked a hyena, chasing him right around the zebra family. The zebras moved on again with the hyenas following. Now and then a hyena managed to get very near to the family or even in between the zebra, biting at their flanks. The speed still did not increase. By 1907 hours [seven minutes after the hyena first approached the zebras] seven more hyenas had been attracted by the commotion and there were 15 hyenas following the zebras. Suddenly one hyena managed to grab a young zebra while the stallion was chasing another member of the pack. This young zebra fell back a little, and within seconds twelve hyenas converged on it; in 30 seconds they had pulled it down while the rest of the family ran slowly on. More hyenas arrived and the little zebra was completely covered by them. At 1917, ten minutes after the victim had been caught, the last hyena carried off the head and nothing remained on the spot but a dark patch on the grass and some stomach contents. Twenty-five hyenas were involved and the whole process of dismembering took exactly seven minutes."

In another case history, Kruuk describes the killing of an adult female zebra. "Within 30 seconds of being grabbed, the mare fell and was immediately covered by hyenas; she died less than one minute after falling. Thirty-eight hyenas ate from the carcass, and at 2040 [fifteen minutes after falling] only the head of the zebra was left; all the rest had been eaten or carried away."

Unlike the lion, which kills its prey with a specialized maneuver—biting the throat long enough to suffocate the animal—hyenas kill through the act of eating—ripping away chunks of flesh, often first from the soft underbelly, causing death through shock, profuse bleeding, and disembowelment. As has been noted in the chapter on lions, popular attitudes toward a carnivore sometimes turn on the precise method of killing the animal uses. The lion's bloodless throttling method is popularly thought to be swift and merciful, whereas disembowelment, the method used not only by hyenas but also by wolves, wild dogs (also called Cape hunting dogs), and jackals, is thought to be a gory, cruelly drawn-out business, mercilessly inflicting severe pain on the hapless prey animal. So while the fact that they hunt at all may elevate hyenas in the eyes of some, their method of disembowelment may detract from any popular admiration.

Even Jane Goodall, who has spent years in the African bush, found the hyena's way of killing repellent. "We were horrified to see, for ourselves, how they ate their prey alive," she and her husband wrote in *Innocent Killers*. "Since that night, we have seen the same gory drama enacted time and time again, for Cape hunting dogs, commonly known as wild dogs, and jackals also kill by this method of rapid disembowelment."

But, being the systematic observer that she is, she discovered, as have a few others such as George Schaller and Hans Kruuk, that this method of killing may in fact be more merciful than that of the lion.

"We still hate to watch it," Van Lawick and Goodall continue in their book, "and yet, though it seems longer at the time, the victim is usually dead within a couple of minutes and undoubtedly in such a severe state of shock that it cannot feel much pain. Indeed, lions, leopards and cheetahs, which have the reputation of being 'clean killers,' often take ten minutes or more to suffocate their victims, and who are we to judge which is the more painful way to die?"

Kruuk timed sixteen incidents in which hyenas attacked wildebeest and found that the average time between the first bite and death of the prey was 6.3 minutes with the range varying from one to thirteen minutes. Zebras are dispatched with similar efficiency and smaller quarry such as gazelles succumb even more quickly.

One curious point noted by virtually all observers of hyenas' killing is that once the predator has fastened its teeth in the flesh of the prey, the doomed animal seems to give up immediately. There is no thrashing struggle with the would-be victim breaking away to escape and lick his wounds. Instead, it looks as if that first bite triggers some kind of shock and the animal collapses, virtually immobile. This obviously speeds up the killing process and could possibly be an evolutionary adaption that, while sacrificing the individual, is good for the individual's species in the long run. If a prey animal escaped with some wounds, it could be at a competitive disadvantage if the damage did not heal perfectly. Weakened or lame, it would still consume food but eventually become an easy mark for another predatory attack. Thus, instead of eating up more food that could be used by its unwounded, species-mates, it concedes early.

Having downed its quarry, more of the hyena's extraordinary physical adaptations come into play. Its shearing teeth, powerful jaws, massive head, and low-slung body are built so as not to waste one bit of the victim's carcass. Not only does the hyena consume the flesh; it eats the bones. Marrow aside, bone is not the indigestible mineral it seems to be. Forty per cent of a bone's weight is collagen, a form of protein that most carnivores are unable to digest but which the hyena makes use of. The hyena's premolars are massive, conical teeth ideal for crushing bone, and farther back in its jaw are the so-called carnassial teeth. Whereas human beings have molars with opposing flat surfaces that crush like a pair of pliers, the hyenas' carnassials shear like a pair of scissors. With its carnassials, a hyena can neatly cut thick pieces of hide or tendon that hold bones together.

Kruuk and other hyena observers have watched hyenas using their various kinds of teeth for specialized eating jobs. Often, to crunch a bone, a hyena will hold it with its forelegs to get it in the proper position. People have even reported seeing hyenas work on a gazelle's head after pushing the horns in the ground to hold it steady.

The splinters of bone that a hyena swallows are digested so completely that the animal's droppings are little more than white powdery calcium. Because the hyenas of a clan often use the same spot for defecating over a long period of time, travelers in the bush frequently find brilliantly white patches of ground here and there around the bush. These are known as hyena latrines.

Unlike the lions, which fight and squabble over a kill, hyenas are remarkably polite when dining. "With 30 or more hyenas frequently eating side by side (I have seen up to 52 eating together)," Kruuk relates, "considering their powerful jaws, such communal feeding could lead to some very nasty fighting, but in practice this does not happen." On large kills, where there is more meat per hyena, the meal is quickly but civilly consumed. On smaller kills, not all the clan can feed at once, and whenever one member wrenches off a leg or other chunk and trots off a distance to eat alone, another hyena or two may chase it for a moment or even try to grab the meat, but, Kruuk says, there is never a real fight. Kruuk says he sometimes tried to take meat away from his pet hyena, Solomon, and while the animal would struggle to keep its hold on the meat, Kruuk could eventually wrench it away and never was bitten.

At a kill, dominance hierarchies do obtain if there is not enough room for all to eat at once. The females have first priority and usually win any struggle with a male. The males, on a small kill, have to wait until the females are full. The cubs come last, and if the carcass does not last long, the cubs go hungry. Unlike the predators of the dog family (wolves, wild dogs, jackals), hyenas do not take meat back to the den for the very young cubs, nor do they regurgitate food for them. But it is not uncommon, if the cubs are big enough to reach a kill site, for a female to stand guard over her feeding cubs, keeping other adults away while the cubs eat.

Of course, it may be unfair to compare hyenas to the dog family, known scientifically as the Canidae, for hyenas are not of this group. Neither are they one of the cats, Felidae, although they are somewhat closer to cats than dogs. Hyenas are in a class by themselves, Hyaenidae, which is most closely related to yet another carnivorous family known as the Viverridae, which includes mongooses and civets, which, to confuse matters further, are often called "civet cats."

Although hyenas are not related to human beings except in the sense that all life is related, the hyena's way of life is similar to what has been suggested for early man. Several anthropologists and wildlife biologists have suggested that we might come to understand our origins better by studying the social carnivores such as hyenas than by studying the apes. Kruuk points out that the hyena's maintenance of territories, even to the point of regularly marking and re-marking borders may have something in common with man's almost universal practice of marking off territories, be they suburban lots, urban neighborhoods, or national fishing limits. Do hyenas and people violate boundaries for the same reasons, for example, when food supplies are dwindling at home? Does the hyena's elaborate greeting ceremony serve a function similar to that of human beings, both of which increase the complexity of the greeting ceremony in proportion to the time separated? Unfortunately, little research has been done to answer these and other such questions that could suggest the origins of human behaviors.

While most of the large carnivores in Africa, and everywhere else in the world, are experiencing a precipitous decline in numbers because of the expansion of human populations, hyenas fare relatively better because they have proven adaptable to man's proximity.

As gastronomic opportunists, some hyenas have become camp followers, trailing about with various African nomadic groups, eating the scraps they throw out or leave behind. The Masai, the tall, proud, and rigidly traditional people often popularly associated with East Africa, are one such tribe, herding cattle and moving from place to place in search of green grass. Most, if not all, groups of Masai have their attendant hyenas. In addition to eating leftovers, the hyenas also act as undertakers. It is a Masai custom, after appropriate services, to place one's dead kinsmen out in the bush some distance from the encampment and allow the hyenas to "recycle" the corpse.

Many settled African villages also have resident hyenas that come at night to the local dump to feed. The city of Harar, in Ethiopia, has become famous for its resident hyenas that not only visit dumps but stroll unmolested through the streets. Not only are the animals tolerated, they are welcomed. People go out to feed them and the hyenas have become quite tame. Harar, thanks to its hyenas, is a very clean city.

The more typical African view of hyenas, however, is that they are pests. People who raise cattle and sheep for example, regard hyenas much as American ranchers once regarded the wolf, for hyenas do kill sizable numbers of livestock. As a result, there have been many campaigns to exterminate them. Game control officers have killed many thousands through shooting, trapping, and poisoning. Incredible though it seems, this attitude has extended from farmers' legitimate efforts to protect livestock near ranches to the national parks where the hyenas play nothing more than their natural and desirable role as a major carnivore. A good example is Nairobi National Park in Kenya, one of the most remarkable of all the East African game parks, for within its small area—forty-four square miles compared with Serengeti's almost six thousand— just outside a modern, bustling city of more than six hundred thousand, visitors may see an extraordinary variety and abundance of wildlife. With the exception of elephants, every major animal species popular with tourists lives there. There are no hyenas. There used to be, but some years ago park officials, not liking hyenas and thinking they spoiled the park's beauty, killed them. They shot some, but for the most part, poisoned them by putting strychnine in bait. No doubt some scavenging lions were poisoned too, but people didn't know lions scavenged then. It should be emphasized that this hyena extermination was undertaken by British conservation officials before Kenya became independent. Since independence, in 1963, management of Kenya's parks has come under African control and, despite the fears of those who said this would doom wildlife, the parks and their animals have largely prospered.

Still, an antihyena attitude persists in many quarters. Hans Kruuk notes that even in Serengeti National Park of Tanzania, where wildlife management policies are guided by a staff of highly educated European and American scientists at the famed Serengeti Research Institute, hyenas were shot on sight until just before Kruuk arrived at the institute to begin his research. Killing hyenas is now prohibited.

From its powerful jaws and bone-digesting stomach to its skill at big game hunting, the hyena is clearly one of the most supremely adapted animals in Africa. Whether it is functioning as the major predator in an ecosystem or as the unofficial mascot and sanitation corps of a town, the hyena is obviously capable of a variety of life styles. Through predation it removes the weaker and slower members of its prey populations, improving the fitness of that species as a group. The hyena is intelligent and maintains a harmonious and peaceable family life. If none of these facts makes the hyena worth admiring, how about the fact that by yielding its kills, it helps support lions?

Where is Walt Disney, now that we need him?

9

BEARS

At one time or another in almost every Westerner's childhood he or she is given a teddy bear. In many cases, this stuffed toy is used as a "security blanket," as something to hug, often providing a child psychological comfort while he is falling asleep. Other animals are similarly honored with stuffed likenesses, but the bear, especially a brown bear, is the runaway favorite.

Bears, especially the brown bears (which include grizzlies), are the single most dangerous carnivores man has had to live with in North America, and, except for grizzlies, in Europe since lions were exterminated there centuries ago. Wolves, as we have seen, have never been real threats to man. Alligators have been largely confined to sparsely populated subtropical regions and, again, seldom a hazard. Cougars almost never attack people. About the only real contenders for the title of Western man's most dangerous animal enemy are poisonous snakes. But snakes cannot begin to rival brown bears as creatures of cunning, power, and ferocity. One seldom hears tales of bear attacks any more but the old settlers and frontier hunters told them often and with the full complement of adjectives such as "vicious," "savage," "hideous," and "terrifying."

Such a fearsome view of the fat, furry bear may seem curiously unenlightened and erroneous to someone raised with a teddy bear under the arm, or to someone whose images of bears includes that of the dignified Smokey Bear, or someone who immediately thinks of such "lovable" creatures as pandas and koalas, neither of which is a bear. (Pandas, from Asia, are most closely related to raccoons; koalas, from Australia, are marsupials like opossums and kangaroos.)

But that fearsome view of the bear is one that prevailed in the United States back when people knew bears better, back in frontier days when a settler going on a bear hunt was risking limb if not life, back when an Indian's bear-claw necklace was an unmistakable badge of courage and skill.

As bears began to disappear from settled regions, the popular image of bears mellowed and Rudyard Kipling could put a wise, slightly bumbling old Baloo in his *Jungle Book,* (1894–95). Incidentally, Baloo, who taught Mowgli the "law of the jungle," is probably the direct ancestor of Smokey Bear who teaches the rules of the forest to modern human beings who venture into it. By 1903 the bear must have pretty well lost its reputation as a ferocious killer, for that was the year the teddy bear was invented. It became an instant commercial success. The story book character that children once called "Bruin" became "Teddy." The stuffed toy, produced by what was to become the Ideal Toy Company, was originally marketed as "Teddy's bear," a reference to a widely published photograph of President Theodore Roosevelt in hunting garb with a spared bear cub playing beside him. The teddy bear was intended as a doll for boys. Children, or at least the adults who bought them, took to the toothless and clawless teddy bear in huge numbers and have continued to do so.

For the vast majority of Americans in this century, the only contact with real bears has been at the zoo, at the circus, where they are made to rear up on their hind legs and "dance" in circles, or at the dank bear pits that once were a fixture of small towns across the United States. In those miserable enclosures, now closed or improved into modern zoos, hundreds of diseased and malnourished bears paced sullenly back and forth or slept as millions of visitors peered down at them and formed their ideas about bears.

The idea that live bears were just overgrown teddy bears was, in fact, once largely accepted by the National Park Service, which, in the days before the parks were so crowded, encouraged the public to visit the parks for the fun of feeding the bears. Publicity photographs frequently featured bears begging for food from cars stopped along the road and even from grinning tourists standing outside their cars offering some morsel, and confident that the bear could distinguish between the bread crust and the fingers holding it. As more and more people fed the bears, more and more bears came to expect handouts from anyone who happened along. Visitors who came to a park hoping to see bears living naturally soon had little better than a zoo-goer's impression of these beasts. Some park visitors who didn't offer food found that the disappointed bears became angry or tried to claw their way through the car window. And, inevitably, people were bitten and clawed and sometimes severely mauled and even killed. Eventually, the Park Service, realizing its mistake, prohibited the feeding of bears, but the habit had been established both among the bears and the tourists. Although more than a hundred people a year are hurt by bears in American national parks, many visitors go on thinking of them as cute, cuddly comics.

In the majority of injury cases, the attacker is the black bear, which, although smaller and less aggressive than the brown bears such as grizzlies, is still not a beast with which to trifle. A few years ago a young girl was killed and partly eaten in Michigan by a black bear. In 1963 the wild blueberry crop in Alaska was poor, and as a consequence a number of black bears that normally relied on the berries for their food at that time of year had to seek other food. That summer at least four people in Alaska were attacked by black bears. A fisherman was dragged from his sleeping bag and severely mauled. A canoeist who had put ashore was chased up a tree and the bear followed, biting his feet and legs. A hunter was dragged from a tent and bitten and clawed in the chest. All three bears in these incidents were killed or driven off by onlookers before their victims sustained fatal injuries. But in the fourth attack that summer, a miner was killed and partially eaten. Alaskan wildlife officials believed these incidents were not simply the results of accidental encounters. They believed the black bears were hungry and wanted to eat the people. Although these incidents were reported in some newspapers, they did not impress themselves on the public mind enough to change the image of bears as cute and friendly.

In the summer of 1967, however, the placid public image in the United States of the lovable bear was shattered by widely reproduced newspaper accounts of an extraordinary coincidence in Glacier National Park in Montana. During a single night bears attacked and killed two young women camping there. The attacks took place some twenty miles apart and were the work of two grizzly bears. In both cases the women, both nineteen years old, were in sleeping bags with a group of other campers. The grizzlies bit and clawed at them, dragged the screaming victims some distance away and, when the women were dead, left them and disappeared into the forest. The killings prompted an outcry from some people for the extermination of grizzly bears. Such marauding beasts, some editorials intoned, must not be allowed to interfere with the rights of people to camp in the forests. Grizzlies have no place in our society. Shoot or poison them all or, as some of the more humane commentators suggested, trap them and remove them to some remote place where they can't hurt people. The outcry fitted well with the goals of local stockmen who had complained for years that grizzlies were eating up their cattle and sheep. A long-smoldering dispute over how to manage the thousand or so grizzlies that live mostly in and around the national parks of Montana, Idaho, and Wyoming suddenly burst into a national controversy.

Park rangers in Glacier found and shot two bears thought to have been the guilty parties and shot two more for good measure. A few people argued that there wasn't much point in having a wilderness if you were going to remove the wild things from it. The public furor finally died down and teddy bears continued to be big sellers.

In 1976 the incident was recalled in a movie and a paperback novel based on the movie, by Will Collins, both entitled *Grizzly* and both advertised as presenting "eighteen feet of gut-crunching, man-eating terror!" Neither movie nor book was an artistic triumph, but they illustrated something about the way in which people think of bears. An ordinary bear, the writers and producers felt, wouldn't be believable as an object of terror. Even though grizzlies are large bears and can weigh as much as a thousand pounds and be six feet in length, the makers of *Grizzly* had their bear weight *two* thousand pounds and "stand" *eighteen* feet tall. Like the gorilla, the bear had been transformed by fiction writers into a two-legged beast when, in fact, neither usually walks on two legs and neither would enter battle in such an unsteady posture. The idea that bears fight reared up on their hind legs probably can be traced to the medieval tradition of training bears to pretend to box while standing on their hind legs and to taxidermists who mount their client's kills in this pose. Eventually, the cinematic grizzly, who repeatedly kills campers, is pursued by squadrons of policemen on foot and in helicopters. He is killed and the forest is once again made safe for people.

In real life, the grizzly (*Ursus horribilis*), the most dangerous of the brown bears, has not fared much better in its competition with people. Although no more than a thousand grizzly bears are estimated to live in the conterminous forty-eight states, there once were probably a million and a half all over the American West. Largely because of the spread of agriculture and hunting, they were exterminated from most of the Plains states by 1870. In the first third of this century most other western states saw the last of their grizzlies die out. Today the only sizable populations outside of Canada and Alaska, where they remain numerous, are in Montana,

Wyoming, and Idaho, clustered around Glacier and Yellowstone national parks.

The thousand grizzlies that survive in and around these refuges have become the focus of one of the most heated wildlife controversies the United States has seen. The controversy, a clash between park officials and wildlife biologists over the status of the animals in Yellowstone, is reminiscent of that involving the overcrowded elephants of Tsavo National Park in Kenya. Both debates represent a collision between established wildlife management policies and the efforts of scientists to gain new information that could change those policies. In a nutshell, the grizzly disagreement is this: The national park people say there are probably as many grizzlies in Yellowstone as the park can support, given the territorial needs of grizzlies (several square miles apiece). They say the population has been growing and as of 1974 stood between 319 and 364 animals. Outside scientists, led initially by the zoological team of Frank and John Craighead, whose work has been the subject of several television specials, studied the grizzlies in Yellowstone and maintained that the population was falling. The Craigheads said that in 1974 there may have been as few as 82 bears in the park. They added that the grizzly may become extinct in Yellowstone by the end of the century. So Yellowstone is either full of grizzlies—or on the verge of seeing them die out altogether.

Because a certain number of grizzlies that harass people are shot or captured and sold to zoos every year by park officials, wildlife preservationists contend that park officials are covering up the effect of these bear losses by insisting that Yellowstone has lots of grizzlies. Some preservationists believe there is actually an extermination campaign underway as part of a quiet effort to meet the demands that arose out of the 1967 killings in Glacier National Park. They argue that park officials are more interested in accommodating growing numbers of human visitors (protecting them from bear attacks by removing the bears) than they are in protecting the wildlife of the park. Although some of the preservationists' assertions are extreme, many authorities agree that American national parks face a fundamental problem in reconciling the mounting demand for public use of parkland with the fact that such use may destroy the very things for which the parks were established.

On the other side of the grizzly controversy, the park officials say that the Craigheads were wrong in some of their assumptions about grizzly behavior and population dynamics and that, in any event, the rate at which bears are removed is declining and approaching zero.

Central to the debate is a behavioral transformation the grizzlies underwent many years ago. They became addicted to garbage. For thousands of years grizzlies hunted and gathered their own highly varied diet of plants and animals. But when pockets of civilization and its inevitable garbage dumps were established within Yellowstone and Glacier parks to serve tourists, the grizzlies learned that they no longer had to make a living the old hard way. Hundreds of bears converged on the dumps and learned to pull the plastic wrappers off moldy loaves of bread and to lick the inside of a can without cutting their snouts. The biggest carnivore to roam the wilds of America since prehistoric times became a mustard-smeared tramp, happily rummaging among the putrid heaps of restaurant and campground refuse. The half dozen garbage dumps in Yellowstone became prime attractions for tourists wanting to see grizzlies.

Most of the Craigheads' studies, beginning in 1959, were

on the garbage-eating grizzlies because they were easy to find. They were ear-tagged, released and some were tracked with the aid of radio collars to learn their movements. It was on the basis of the activities of these animals that the scientists reached their estimates of Yellowstone's grizzly population. The National Park Service contended that this was an under-estimate because it did not count "back country grizzlies," animals that never came to the dumps. The Craigheads said there wasn't much evidence that such bears existed.

When the two young women were killed in Glacier Park in 1967 and it was learned that one had been camping just a few hundred yards from a dump frequented by grizzlies, there arose an immediate cry to close the dumps. The park officials wanted to do so quickly. The Craigheads said they should be closed gradually over a period of ten years so that all the bears would not suddenly be sent hungrily scrambling into the nearest campsites for something to eat. Park officials, annoyed at the Craigheads for disagreeing with them, insisted that they submit their reports for official approval before publishing them. The scientists refused, and their research was ended with no love lost on either side.

The dumps were closed over a two-year period and, depending on whose figures and opinions you accept, the number of grizzly attacks on people either did or did not increase. All agree, however, that the number of grizzly removals increased immediately afterward. Followers of the Craigheads say the population of bears has been declining because the removals exceed the number born each year. Park officials say the population is growing because, now that the bears are no longer crowded around the dumps, natural birth control methods induced by crowding (a phenomenon common in many animals) have relaxed and more females are reproducing.

At one point the arguments became so acrimonious that Interior Secretary Rogers C. B. Morton asked the National Academy of Sciences to make an independent evaluation. The report found fault with both sides but leaned somewhat toward the Craigheads's conclusions.

One measure of the difficulties inherent in the controversy over grizzly population status is indicated by the fact that the United States Government has been changing its mind repeatedly for ten years about whether or not the species is endangered. Grizzly bears were put on the official Endangered Species List in 1966 and then removed from the list in 1968 because a thriving hunting industry in Alaska insisted there were plenty there. In 1973 the law was changed to permit calling a species endangered in one region even if it is plentiful elsewhere, and in 1975 the grizzly was declared not "endangered" but "threatened," a less severe predicament in the vocabulary of the U. S. Fish and Wildlife Service, which keeps the List. Then in 1976 it was decided that the grizzly was endangered after all.

Throughout the grizzly debate, which is likely to continue for some years, one major concern of everyone has been to prevent bears from attacking people. For all the scientific studies, publicity, investigative commissions, and such, the death toll was never all that great, certainly not by comparison with other causes of accidental death. In fact, most years go by with no deaths at all from grizzlies. In 1975, the latest year for which figures are available, grizzlies injured five people in Glacier National Park and two in Yellowstone. There were no deaths. The number of grizzly attacks is a tiny fraction of those perpetrated by black bears, which, though less aggressive, are far more numerous.

Even so, it is clear that much of the public is unwilling to tolerate even the slightest evidence that there should be dangerous wild animals lurking about. If there should be a repeat of the killings of 1967, it is virtually certain that a wide demand for the extermination or removal of the grizzlies will rise again. And it is not unlikely that the demands will be accepted and acted upon.

Garbage-eating grizzlies and begging black bears are popular. They fit the stereotype some people entertain of the harmless, lovable bear. But when people take it upon themselves to enter bear country and encounter a bear behaving like a dangerous meat-eater, then they argue that it is not they who must go but the bears. These people want real bears toothless and clawless, just like the teddy bear. As long as bears are playful and seemingly harmless, they are liked. But, let a bear behave like the wild carnivore it truly is and there is no longer room for it in some people's world.

10

CROCODILES

Throughout history reptiles have repeatedly supplied moralists, authors of adventure stories, and other writers with personifications of evil. These cold-blooded creatures, scaly beasts that crawl on their bellies and lurk about in low places, seldom inspire admiration or affection. The evilest of the reptiles, the most dangerous and loathsome, by some accounts, is the crocodile, a beast that inhabits many waterways of Africa, Asia, and Latin America.

A possible early example of this tradition may have been

the serpent in the Garden of Eden. Although Eve's tempter—
Satan incarnate—is usually depicted as a snake, the word
"serpent" was an ancient umbrella term for any animal that
creeps, hisses, or stings, including the crocodile. Leviathan,
who appears several times in the Bible, is usually understood
to be a sea serpent representing the devil, but its descriptions,
as in Job 41:1–34, suggest the aquatic Nile crocodile, the
only real species known to have existed in the ancient Middle
East that resembles the description in Job and the only model
on which to pattern the Leviathan myth. Although the term
"leviathan" is sometimes thought to mean whale, it is clear
from the descriptions in the Bible that a whale is a less likely
leviathan than a crocodile. Job's leviathan spewed flames
from its mouth and smoke from its nostrils and was impervi-
ous to swords and spears. Though smoke and fire fit neither
whale nor crocodile, leviathan's tough exterior seems more
like a crocodile's than a whale's.

The leviathan has come down to us in somewhat more recognizable form as the fire-breathing dragon of the Middle
Ages, which none but the virtuous could slay. Again the
beast, though often depicted with some mythical characteristics, seems obviously patterned after the crocodile, the
biggest of which can be from fifteen to twenty feet long and
weigh up to a full ton.

While the hyena and baboon are disliked, by man, it is
chiefly because they are thought repulsive in form or habit;
they have seldom been regarded as inherently evil. The crocodile is similarly disliked and thought repulsive, but there has
also been good reason to think of them as malevolent for they
commonly kill and eat human beings—particularly innocent
mothers and children whose traditional jobs in many societies
include going down to a crocodile-infested river to fetch water

or do the laundry. And, too, crocodiles are relics of a time millions of years ago in the dim, dismal past of the Age of Dinosaurs, when, in the popular view, the highest forms of life were brutes and bullies.

The identification of the crocodile with evil is evident in this excerpt from an 1888 issue of *Sunday School Advocate,* a weekly take-home publication for children. Beneath a drawing of a crocodile devouring a small child there is a long narrative about how crocodiles and their cousins the alligators snatch unwary children who can escape only if they remember to gouge out the beast's eyes. The graphic description ends with a moral: "But though this scaly monster does not haunt the rivers of the North, yet there is another great dragon ever prowling about our streets and watching our homes, seeking whom it may devour. It is more terrible than the alligator. Its jaws are mighty to crush and destroy all its captives. The name of this monster is SIN! Children, beware of it; keep out of its way, and you will be safe."

Although the crocodile-as-evil idea may have lost its power these days, this is probably due not to the fact that we understand crocodiles better but that we understand evil less. The crocodile's evil connections today are reduced to little more than *Flash Gordon* reruns, in which they infest pits the bad guys force the good guys into.

In the case of the crocodile, Western attitudes are not so different from those of Africans who live close to the big reptiles. Africans don't like them any more than Americans or Europeans do. However, some tribes do consider crocodiles to be executors of justice. If someone from these tribes is eaten by a crocodile, it is taken as good evidence that the person was evil and deserving of the fate.

Although there could be many theories about why the croc-

odile is perceived as evil, Alistair Graham, a former Kenya
game warden and professional wildlife consultant, makes a
particularly persuasive case that it stems from the fact that
crocodiles do eat people—more so probably than any other
predator. The traditional taboo against cannibalism is so
strong, Graham says, that we do not easily distinguish between people eating people and animals eating people. Graham delves into this in his book *Eyelids of Morning* (1973),
a study of "the mingled destinies of crocodiles and men"
based on his research in Kenya. "It is around the matter of
cannibalism that the symbolism of crocodiles twists and
writhes," Graham writes. "To be eaten by a croc is to be consumed forever by evil. One forfeits all hope of immortality.
One's soul is irrevocably Satan's, one's body is dung."

In the 1920s and 1930s there was a crocodile in Lake Vic-
toria that used to come ashore near Entebbe, Uganda, whenever called by local villagers. The huge beast, named Lutembe, was only semitame, for on occasion he would eat
people. Over the years, however, Lutembe came to be regarded as an arbiter of justice who bit or ate only people who
were guilty of some crime. Suspected thieves were brought before Lutembe for judgment. There was, of course, no possibility of appeal from a conviction by Lutembe.

Because of its reputation, the crocodile has suffered greatly
at the hands of human beings. Africans killed them whenever
possible, which was not often before Europeans introduced
the high-powered rifle. With the white people came the beginning of the decline of crocodiles. Once common all over
Africa, they are now reduced to a fraction of their range of a
century ago. Although apparently in little danger of extinction, crocodiles are gone or nearly gone from most of the
densely inhabited parts of Africa. Where they remain, there

are few of the large twelve- to fifteen-footers. The biggest of these slow-growing animals have either been killed for their skins, which have been fashioned into handbags or shoes, or simply left to rot on the shore. Consequently, it is a rare crocodile these days, except in the remotest parts of Africa, that grows much beyond its young adulthood length of six to nine feet.

A major hero in the war on the Nile crocodile was Captain Charles Pitman, Uganda's first game warden who, in the first third of this century, waged a seventeen-year campaign to exterminate the species. Although Pitman did not believe in the croc's innate evilness—"my own summing up of the crocodile's character is that this foul beast is a typical bully and a great coward"—he used every method at his disposal to finish them off, from poisoning, shooting, and trapping to offering bounties for crocodile eggs. So successful were Pitman and his successors that Uganda today is largely free of crocodiles, except for a modest population in the Kabalega Falls National Park.

As if the unpopularity of the living crocodile were not enough, the popularity of products made from crocodile skin gave a further boost to the killing rate. Crocodiles had long been considered vermin by Africans and European game departments alike, but with the demand for crocodile-skin wallets, shoes, and handbags, they were turned into a valuable natural resource, at least as long as the fad lasted. The demand continues, of course, but at a much reduced pace.

Today crocodiles are no longer officially classed as vermin. They have been promoted to "game," meaning that hunting them is thought worthy of regulation by the licensing of hunters and the setting of kill limits. Illegal hunting continues, however, because game wardens cannot catch every poacher.

In addition, there are not enough wardens to patrol every lake where people wish to bathe in safety or at every riverbank where parents take steps to insure that their children are not eaten when they fetch water.

Whether or not Westerners think of the crocodile as an evil creature, many think of it as a sluggish hulk of a reptile that spends most of its life sprawled motionless in the sun. Though crocs do indeed bask for long periods in the sun, recent studies have revealed that they live far more complex lives than might appear to be the case. We now know, for example, that crocodiles hunt co-operatively, share their food, live in social groups with dominance hierarchies, and are surprisingly attentive and conscientious parents.

Perhaps the most unexpected finding involves the parental behavior of these huge reptiles. Crocodiles become sexually mature at around twelve to fifteen years of age, by which time they are seven or eight feet long. During the one time each year in which crocodiles are sexually active, they are monogamous. Male and female engage in a courtship ritual in which they rub jaws and raise and lower their heads. They also hold their jaws open toward one another but do not bite. Then, after two or three days of this, the couple slide into the water and copulate. Five months later, the female is ready to lay her eggs and begins to choose a nest site. If she has reproduced before, she will most likely use the same nest site and continue to do so year after year. The female digs a hole in the ground and buries the eggs (sixteen to eighty) twelve to eighteen inches deep. For the next three months the mother remains by the nest, charging any intruders and eating nothing. The father, although he goes off now and then to feed, spends most of his time near the nest patrolling the shore to keep away any subdominant males.

When the eggs hatch, the baby crocs, still buried under at least a foot of dirt begin peeping loudly. On hearing this sound, the mother scratches with her front legs and jaws to remove the dirt cover. Then she picks up the hatchlings—each weighing about four ounces—one by one with her teeth until all are in her mouth, and gently carries them down to the water. This remarkable event, which has been documented by Anthony C. Pooley, a South African specialist in crocodile reproductive behavior, is made possible by the crocodile's ability to depress its tongue to create a pouch in the back of its mouth. Pooley says the mother gets each baby into the pouch by carefully grasping it in her teeth and "flipping" it back into the pouch.

When she gets into the shallow water, the mother opens her mouth and swings her head from side to side, releasing the young. They are washed free of dirt and sand as they swim to shore and mill about at the water's edge, emitting pulsating chirps. The sounds attract the male whose approach is greeted by the female with what Pooley calls "a low warble."

Pooley has found that if the female is, for some reason, unable to reach the buried hatchlings, the father will dig them out and carry the young to the water. Pooley has even observed males gathering unhatched eggs in their mouths and gently rolling them between his tongue and palate to break the shells and free the babies.

"If one remembers that the weight ratio of the adult to the hatchling can be as much as 4,000 to one," Pooley and Carl Gans, a University of Michigan zoologist, wrote in an article in *Scientific American* (April 1976), "both the male's grasping action and his palpation of the egg are demonstrations of a spectacular oral sensitivity and muscular control. The same jaws that can crush the femur [thigh bone] of a Cape buffalo can pick up an egg without harming the little crocodile inside."

This is hardly the expected behavior of a hulking brute with steel-trap jaws.

Instead of abandoning the hatchlings, both parents remain close to the young for nearly twelve weeks, protecting them from predators. If one of the babies gets into trouble, it emits a high-pitched distress call and its siblings join in. The parent crocodiles immediately move toward the sound to rescue or protect the babies.

After this twelve-week period, which Pooley and Gans call the "crèche phase," the young crocodiles, which are born able to catch insects, snails, and small fish, gradually disperse to live on their own. Up until the time of dispersal, when young crocs see a larger crocodile moving toward them—they move toward it. But after the dispersal, the reaction reverses and the young flee from an approaching larger crocodile. To protect themselves from bullying, bigger crocodiles, the juveniles dig tunnels into the riverbank. Sometimes several young crocodiles work together, digging as much as ten feet into the bank. Pooley and Gans have found that the tunnels may be used for five years by the same animals.

As the crocs grow, they graduate to larger and larger prey species, from insects, to frogs, to birds, to antelopes. Throughout their lives they eat fish of the appropriate size. Contrary to popular impressions, Pooley and Gans say, crocodiles do not simply sit and wait for food to come to them; they actively cruise the waters, hunting. After a few powerful sweeps of the tail and propelled by four strong legs, a crocodile can leap entirely out of the water, covering in a single bound a distance several times its own body length. If the riverbank is steep, say Pooley and Gans, the crocodile "appears to vault straight out of the water." If it needs to run after its prey on land, it does so, holding its body well off the ground and running as fast as a human being can.

Not all the crocodile's hunting is a matter of pure power and speed. Some of it is cunning. It has been seen eying a nest of bird hatchlings in the reeds along the water's edge and then arching its tail around to bend the reeds, spilling the nestlings into the water where they are, so to speak, sitting ducks. A crocodile's tail is also used to herd fish into its jaws. To do this, the crocodile swims along the riverbank with its tail curved toward the shore. The tip of the tail riffles the water gently, keeping the fish moving ahead of it until the crocodile suddenly swings its head toward the fish and snaps them up. "The Nile crocodile is so agile," Pooley and Gans say, "that a fast reverse sweep will even intercept fish that are trying to escape by jumping over its back."

Crocodiles sometimes co-operate at feeding time by helping each other tear up a carcass into bite-sized chunks. Ordinarily a crocodile does this itself by clamping its jaws on one part of the dead animal and spinning around until the mouthful tears off. When this doesn't work, the crocodile will move the carcass toward another crocodile. The second animal bites the carcass and holds on while the first rotates once again. Each crocodile keeps what it tears off. Pooley says he has also seen two crocodiles walking overland, carrying an antelope carcass between them. Still another form of co-operation among crocs occurs just after the rainy season begins, sending heavy currents downstream laden with fish. Where these streams enter a wide natural depression, or pan, the crocodiles arrange themselves in a semicircle before the spreading flow, trapping the fish where they can be caught easily. Each crocodile stays in its place and there is seldom quarreling over who gets which fish.

The crocodile, like all creatures, is simply living in the way to which it is suited. And it happens to be well suited to a wide variety of habitats and foods, which accounts for its success. It has survived as a recognizably distinct animal form for one hundred seventy million years, vastly longer than practically every other animal of comparable size. Most of today's mammals, for example, cannot be traced back more than five to ten million years.

"It is," Pooley and Gans wrote of the crocodile, "nonetheless in danger of extinction, not because it has failed to find its place in nature but because it is the prey of human hidehunters. One might not want to have a crocodile in every water hole, but it will be a sad day when a few stuffed or pickled specimens are all that remain of these splendid animals."

Perhaps this chapter should end on that climactic note, for we have turned evil serpents into "splendid animals." But, though that expression of qualified admiration might be useful in balancing the croc's image, it could prompt the beginning of an overreaction or lead to the idea that crocodiles should be protected because they are "splendid." I am not talking about whether or not we should conserve crocodiles in their natural habitat. We should, and more will be said of this in the last chapter. But it is no better to think of crocs as good than it is to think of them as bad. An animal is not worthy of protection because our current value system makes it a hero any more than an animal should be exterminated when we regard it as a villain.

Crocodiles are "splendid" all right, but so are snakes, hummingbirds, cockroaches, lions, rats, and human beings.

11

WHALES and DOLPHINS

The latest group of animals to be added to the misunderstood menagerie belong to the order Cetacea, i.e., whales, particularly the small whales known as dolphins, or porpoises, that are popularly believed to be highly intelligent and, according to some admirers, to rival human beings in intellectual ability.

Few who have visited one of the "Marinelands" or "Sea Worlds" to see the dolphins have not marveled at what seems to be the intelligence and playfulness of these graceful, air-breathing, aquatic mammals as they execute an astounding variety of tricks. Who can fail to wonder what goes on inside their human-sized brains when they seem so obviously to enjoy human company? And who has listened to the recordings of long, mournful whale songs and not imagined that they were oratorios or at least intelligent messages resounding through the ocean depths? Dolphins have even penetrated popular culture through a now-defunct television series called "Flipper," which was to dolphins what "Lassie" was to dogs, and a Mike Nichols film called *The Day of the Dolphin* (1974).

As entertaining and even as inspiring as the merchandising of these creatures may make them out to be, there is no good evidence that dolphins are any more intelligent than a number of the smarter terrestrial mammals such as dogs and monkeys. Less is known about the larger whales, but there is no reason to believe that they are any more intelligent than their smaller cousins. The popular fascination with these creatures appears to derive less from objective evidence than from the traditional mysteriousness of the sea.

It is attractive and even humbling to think of the oceans, truly one of the least explored regions of our planet, as a world of highly intelligent creatures, perhaps rivaling ourselves, who live in a utopian society in peace and brotherly love. Lacking hands, the whales and dolphins may not have produced the kind of technological artifacts by which we judge social advancement, but, this school of thought holds, because they have very large brains with deep convolutions like ours and because they engage in such rich vocal communication, it is argued that they must be very advanced socially. That this is not immediately apparent is said to be our fault, not theirs.

Anyone who has not made contact with the large and growing popular cult of the cetaceans may find such sentiments a bit much, but they are remarkably widely held, especially by younger people. The cult is even encouraged by a number of general conservation organizations, not just those activist groups dedicated exclusively to the cetaceans. Not long ago, for example, the Sierra Club published a book, compiled by Joan McIntyre, called *Mind in the Waters* (1974), a title drawn from the ancient belief that dolphins were the embodiment of the mind of God moving through the waters. On the book's cover, it says, "We are beginning to discern the outline of another mind on the planet—a mind anatomically like ours but profoundly different."

Although dolphins have fascinated people for thousands of years (the ancient Greeks, who called the beasts *delphys*, linking it to *delphis*, which means "womb," and to the city of Delphi, where the "navel" of the world was alleged to be, considered it the embodiment of the sea's vital power), the modern cult of the dolphin is traceable to one man, John Lilly, M.D., who specializes in neurology and began studying dolphins in 1955. Lilly has written two books, *Man and Dolphin* (1961) and *The Mind of the Dolphin: A Non-Human Intelligence* (1967), for the layman. They recount man's long historical fascination with dolphins, the many tales of how dolphins push drowning swimmers to shore, and numerous instances in which captive dolphins seem to have learned tricks with astonishing rapidity or to have established an uncanny rapport with their human captors.

Lilly, once a respected scientist, began his first book with the assertion that within ten or twenty years man would establish communication with another nonhuman species "possibly extraterrestrial, probably marine; but definitely highly intelligent, perhaps even intellectual." This achievement, Lilly said, would reveal "ideas, philosophies, ways and means not previously conceived by the minds of men." Lilly himself hoped to be the scientist to make the breakthrough and he devoted years to working with captive dolphins in Miami and the Virgin Islands. Although the time allotted for Lilly's forecast to come true has not yet expired, Lilly himself has long since given up the effort and now no longer does research on dolphins. He quit not because he lost interest but because he concluded in his own mind not only that dolphins were highly intelligent but that they possessed civil rights that he should not violate by experimenting on them. Lilly today has largely abandoned science and now devotes himself to spiritual and mystical pursuits.

His books, however, are still in print and have stimulated a number of similar books by other authors and prompted scores of people who have access to dolphins to take up the goal of establishing communication with them.

By extension, Lilly's opinions about dolphins have been applied to the larger whales. Dolphins are just small whales, people say, correctly. Therefore, they add without evidence, whales must be as smart as dolphins. Some even think they must be smarter. The brain of the sperm whale, for example, is five times the size of that of the Atlantic bottle-nosed dolphin, the species most familiar to the American public.

What is it that has persuaded so many people of the dolphin's intelligence? Two things.

One is this business of the brain. A dolphin's brain is about the same size as a human brain and has deeply folded convolutions all over the surface. These convolutions are known to be a way of increasing the area covered by the outermost layer of the brain, the cerebral cortex, which, in human beings, does the thinking that we normally consider as distinctively human.

However, although it is often claimed that the dolphin's brain is the same shape as a human brain, they are really quite different. The dolphin brain, in fact, has a very different shape from that of most other mammals, being more nearly globular while others are flattened on the lower side. It also possesses a number of structures that neurologists are at a loss to explain. Detailed examinations of dolphin and whale brains also show them to have, compared with human beings, a lower density of individual nerve cells. Thus, although the brain may be as big over all, it has fewer nerve cells in it. Also, the dolphin's cortex is thinner in relation to the rest of the brain than are those of human beings.

It is possible that in the larger whales these factors might be more than compensated for by the enormously larger brain. Whereas the brains of human beings and dolphins weigh around 1,600 to 1,700 grams, the sperm whale's brain is a hefty 9,200 grams, or just over twenty pounds. A sperm whale, however, can weigh a hundred tons. While brain size is roughly correlated with intelligence, one must take into account the proportions of the body in which the brain resides. For example, an elephant's brain is about four times as big as a human being's, but its body is about fifty times as big. It is necessary to take body size into account because the brain must accept and process information from nerves throughout the body, and the more nerves there are in the body, the bigger the brain must be to handle them.

Another consideration in comparing brains is whether or not the species exercise any nonintellectual functions that would still require lots of brain tissue. Dolphins and whales do. They employ a kind of sonar by which they send out high-pitched sounds, listen for the pattern of their echoes, and, mentally, construct a kind of acoustic picture of the world around them. Thus they can be blindfolded and still swim unerringly through an obstacle course toward a ball floating on the surface a hundred yards away. Certainly it takes a fair amount of brain capacity to do this. It may be, for example, that the dolphin's brain gives it an acoustic picture of its world every bit as real and believable as the visual picture constructed in human brains on the basis of reflected light waves.

The second thing that has persuaded so many of the people of the dolphin's intelligence is its behavior under certain circumstances, particularly its ability to mimic sounds and actions. They are much better at it than parrots.

Dolphins can be trained to perform tricks at least as easily as seals, dogs, and other animals and usually much more easily, for they need fewer demonstrations to get the idea and less reward as an incentive. Lilly's books and all the others are full of accounts of how dolphins pick up tricks after being shown only once what to do. What makes this behavior interesting is that it is not just formal tricks that they pick up. Dolphins will imitate almost anything they see, if it is possible, given their anatomical structure. There is an account, for example, of an Atlantic bottle-nosed dolphin put in a tank with a Pacific dolphin of another species. It happens that the Pacific dolphin makes spinning leaps out of the water whereas the Atlantic species makes nonspinning leaps. But when the Atlantic beast saw its Pacific cousin do a spinning leap, it immediately imitated the stunt, succeeding on the first try. There is an even more remarkable example involving Indian Ocean dolphins, which are of yet another species. These animals were put in the same tank as Cape fur seals, and before long they were imitating the seals' movements for swimming, sleeping, and even sex. Divers who maintain the insides of the huge tanks at the oceanaria are generally familiar with the dolphin's inveterate mimicry—it is not uncommon, apparently, to have dolphins imitating the divers' movements, for example, in scraping algae off the glass viewing windows or vacuuming the tank bottom. The mimics even time the release of their own air bubbles from their blowholes to match the breathing rate of the divers. (Dolphins normally do not exhale underwater.)

Perhaps even more astonishing is the dolphin's ability to mimic sounds such as words spoken by people nearby. The accounts of this indicate that the words may be either training commands or simply random human conversation that the dolphin overhears. Some of the dolphin "talk" is in a high-pitched, Donald Ducklike voice that makes sense only when tape recorded and played back at a slower speed. In rare instances, the dolphins have learned to slow down their repetition to the normal human voice range.

It was this ability of dolphins on which Lilly hoped to capitalize by teaching them to speak English. If they have big enough brains and can say words, he reasoned, man should someday be able to converse with them in ordinary spoken English. Despite years of effort at this, neither Lilly nor his colleagues nor others who have made similar attempts have succeeded. There is no evidence that the dolphins are understanding the words beyond the level at which a dog understands words like "sit" or "heel." There is plenty of evidence, on the other hand, that dolphins use sounds to communicate with one another. The sounds are squeaks, clicks, whistles, and assorted other effects that have been recorded and studied for patterns. There *are* patterns and researchers have discerned twenty to thirty distinct signals, the number varying by species. There are probably others yet undiscovered, particularly signals used in the open sea that would be inappropriate in tanks where the studies have been made. Even if the vocabulary of signals is double the number recorded so far, however, it would probably not exceed in size the repertoire of signals used by monkeys and apes.

Much of the popular writing on dolphins has offered their mimicry as evidence that the beasts are extraordinarily intelligent and, in many cases, trying to communicate with human beings. Many dolphin activities are interpreted as the beast trying to teach the trainer something, and it is frequently remarked that sometimes human beings must seem awfully stupid to the dolphins.

There is another explanation that demands less of a leap into the unknown. Richard J. Andrew, a Yale University neurophysiologist and expert on animal communication, has suggested a reason why the dolphin's open-ocean life style might make a knack for vocal mimicry crucial to survival. It would be the same reason birds and many primates also exhibit an ability to mimic their parents, shaping their "accents" to conform to the group's norm. It permits the animals to develop a set of signals unique to their own social group by which they may recognize one another at a distance. Many birds have this ability and it has been shown that there are regional dialects among the calls of a given species. In other words, the blackbirds in one forest have a common set of calls that is different from those of the blackbirds in another forest.

"This faculty would seem to be especially valuable to animals that cruise the open sea at high speeds, repeatedly joining and breaking away from schools of their own species," says Edward O. Wilson, a Harvard zoologist and expert in the evolution of social systems of animals. Wilson wrote the landmark book *Sociobiology* (1975), setting forth the view that the behavior of all social, or group-living, species evolves so as to maximize the survival of the species. He believes that this factor alone would account for the large size of the dolphin brain.

Dolphins are a highly social species whose family group structures resemble those of ungulates. Like the browsing and grazing species, such as antelopes, elephants, and zebras, dolphins "browse" on food (fish) that is patchily distributed throughout their environment. To find enough to eat, dolphins cannot always travel in tightly packed schools, or pods. They must range out of sight of one another, but, in spite of the featureless environment in which they live, later rejoin the other members of their group. There is no way to mark a home base or a path or the border of a territory. Consequently, the only method available for dolphins to rejoin their families is through vocal communication and echo location (sonar). Like the ungulates, dolphins form dominance hierarchies, with an older bull ruling over females and younger males.

One example of dolphin behavior that is often cited as evidence not only of their allegedly superior intelligence but also of their sense of social responsibility, is the rescuing of disabled fellow dolphins. Most often this takes the form of one or more able dolphins swimming under a disabled comrade and lifting it to the surface periodically to breathe. The story is told so often by dolphin partisans that the impression has been created that this is standard behavior. It isn't. The usual response of dolphins to the injury or disablement of one of their number is to abandon the hapless individual and rapidly swim out of the area.

But however rare, the phenomenon is real and cannot be

dismissed. It must be taken as evidence of a sense of altruism. But altruistic behavior among the animals is not limited to dolphins. Many other species try to rescue or protect their disabled kin. African elephants, as we have seen, will struggle for long periods trying to lift a fallen or dead elephant. Mother elephants have been seen shading their young who have collapsed in the open sun; they will stand over the little one for days, ignoring their own needs for food and water. Baboons too will carry a wounded troop member to safety and share food with it. There are similar accounts of African wild dogs behaving altruistically.

"However," Wilson says in *Sociobiology,* "it does not necessarily reflect a higher order of intelligence. By itself, the behavior is not as complicated as, say, nest building by weaver birds or the waggle dance of honeybees. It could well represent an innate, stereotyped response to the distress of companions."

Although many such forms of behavior among animals are termed altruistic, this does not mean the act is entirely selfless. Social animals are better able to survive if their social groups are kept intact and functioning, and any activity that promotes this is to the advantage of the altruistic individuals. Such stereotyped behavior patterns evolve because whenever the rescue of a disabled family member enables the individual to survive, it means that more of the genes shared by the family will be passed on to subsequent generations.

Whether these observations of dolphins apply to the larger whales is only speculation. Very little study of free-living dolphins has been done and even less has been attempted of the larger whales. However, it is known from the recordings of the songs of the humpback whale that this beast, too, makes some use of fixed programs of acoustic messages. Large whales travel even farther apart than dolphins, and their eerie whoops and bellows may well be audible signals by which a whale family, spread over miles of ocean, keeps in contact.

. The songs of the humpbacks are, of course, widely known from commercially available recordings, some of which have been set to symphonic music. To hear these songs, with or without the human enbellishments, is indeed a haunting experience. The songs make it easy to see why Herman Melville said that if God returned to earth it would be in the form of a great whale. But it should not be necessary to impute to the whales any supernatural character or even any great intelligence to justify conserving them. Unfortunately, many of the most ardent save-the-whale promoters do just this. There is much in the whales, large and small, to appreciate on purely objective grounds.

Whales are among the few dwindling species now on the earth for which the major threat is not destruction of habitat but hunting. Furthermore, whale hunting is chiefly done not by people with no alternative food source. Most of the whaling is done by Japan and the Soviet Union where alternatives to whale products for food and commerce are already well known. The Eskimo and other technologically primitive peoples who still hunt whales as they have for centuries represent no threat to whale populations. Dolphins, though not hunted directly in any great numbers, are killed by the thousands by the Pacific tuna industry when they become entangled in a certain type of tuna net and drown because they cannot surface for air. Here again the culprits are from the world's wealthy countries who could afford alternative methods of tuna fishing.

The battle to save whales and dolphins is gaining ground slowly. Much of the popular support it is winning is tied very closely to the idea that cetaceans are sentient beings, members of vast and alien cultures being destroyed on the earth even as billions of dollars are being spent to hunt for microbes on other planets. The enemy, in the eyes of the whale sentimentalists, is the view that cetaceans are just hunks of mobile blubber with great commercial value. It may be that desentimentalizing the appreciation of whales and dolphins would retard the worthy efforts to save them. But that shouldn't be.

12

MURDER

One of the most widespread and deeply held myths about the animal kingdom is that, unlike human beings who murder and maim their own kind, animals seldom, if ever, attack other members of their own species. Everyone knows that the "law of the jungle" permits one species to prey upon another. That is one thing. Many animal lovers also believe there is a contrasting "law of the species" that prevents murder of one's own kind, one's species mates. Unlike other animals, man—it is said—has lost his purity and turned toward outright murder.

Muggings, knifings, shootings, lynchings, riots, massacres, and wars are all said to show man's essential depravity.

Robert Ardrey's book *African Genesis* (1961), contending that man descended from a killer ape, has added to the conventional wisdom of man's innate bloodthirstiness. Konrad Lorenz has pushed the idea still further by asserting that the animals have developed ritual signals that blunt their hostile energies and prevent their harming each other. Man, Lorenz has said, lacks these signals and, therefore, is redder in tooth and claw than the animals.

In its most extreme form, the view is often expressed that man is the only creature that has lost all sense of preservation of the species, the only animal that murders. Sometimes it even comes out sounding as if humans are bestial but animals are humane. This popular concept is, to put it squarely, flatly wrong. It is based mostly on a lack of knowledge about animal behavior and fallacious interpretations of what few facts have been available.

Although the era of intensive, objective observation of wild animal behavior is still young and relatively few species are known well, it has already been found that many animals are far less peaceable within their species than is man. In fact, it is looking more and more as if *Homo sapiens* should be regarded as one of the least violent, least murderous of creatures.

There are now enough detailed behavioral studies of several animal species to reveal that a fair number are intensely murderous and to suggest that a great many more will be found to be the same, as studies in progress are completed and published. There are scientific reports on vicious assaults and murders among lions, cougars, wolves, hyenas, black bears, grizzly bears, macaque monkeys, langur monkeys, black-headed gulls, another bird species known as the brown booby, and many species of insects, such as soldier ants.

"The annals of lethal violence among vertebrate species are beginning to lengthen," writes Edward O. Wilson in his book *Sociobiology*. "Murder is far more common and hence 'normal' in many vertebrate species than in man. I have been impressed how such behavior becomes apparent only when the observation time devoted to a species passes the thousand-hour mark. Only one murder per thousand hours per observer is still a great deal of violence by human standards."

Take the African lion as an example. Out of all the scores of thousands of lions in Africa, George Schaller randomly selected a few prides of, typically, ten to twenty lions each and observed them over an arbitrarily chosen span of time.

Although Schaller studied many hundreds of lions during 2,900 hours of observation over the three years he spent in the Serengeti National Park, he was able to make repeated observations on fewer than two hundred individuals and the vast majority of these were seen only infrequently. Still, Schaller recorded fourteen killings of one lion by another. If we assume that Schaller had all two hundred lions under surveillance for the entire 2,900 hours, which amounts to about a third of a year, the lion murder rate works out to be about two thousand times as high as the murder rate in the United States, which has one of the highest rates of any country in the world. Even if we assume that the fourteen killings Schaller recorded were all that occurred among all 2,400 lions estimated to live in the Serengeti during the full three years he was there, the murder rate still works out to be nineteen times that of the U.S.

Schaller may have happened into the Serengeti at a particularly bad time for lions, a time when there was an unusual level of violence toward one's own kind. But there is no reason to believe so. Because the violence levels were more or less constant over the three years of his study and because he found one pride of lions to be about as violent as another, it is more likely that the level of violence and murder Schaller saw was about normal for lions.

Can it be true that people are so much more peaceable than the seemingly placid lions? Let us assume, as Wilson once suggested, that a scientist from another planet were to visit Earth to make a long-term behavioral study of a few typical earthling families. Let us assume our alien observer is able to watch the goings on in your house without your knowledge and that he has done so for the past three years, the period Schaller observed the lions. Would he have seen one member of your family attacking another violently enough to break bones or draw blood? Would he have seen someone murdered in your house? What about the houses of your friends? Would the alien observer have seen any fathers killing their children? If human beings were as murderous as lions, there should be one or two slayings on the typical suburban block every year and a dozen or two annually on the typical apartment house block.

But what, one might ask, about all the violence reported in the newspapers? It happens in other homes and in other places but those *are* human beings killing one another. In recent years the annual murder rate in the United States has been running at around nine to ten killings for every 100,000 people. Taking the higher figure, then, there is one murder for every 10,000 people. This means that our alien anthropologist would have to observe 10,000 people continuously for a full year to witness a single murder.

The murder rate for the world is unknown, but inasmuch as

every other country but Kuwait and Sri Lanka report lower murder rates than the United States (the British, for example, kill one another only one twentieth as often as do Americans), it is safe to assume that the murder rate for man as a species must be lower than it is in the United States. Thus even the favorable comparison between the American murder rate and that of lions does a disservice to mankind.

But, a "people-are-beasts" proponent might argue at this point, that only man makes war, killing tens of thousands in massive nonstop carnage. Only among human beings could a Hitler arise to direct the slaughter of millions of his own kind.

While it is largely true that animals do not engage in group battles (there are some exceptions, such as wolves), the statistics still do not make man out to be among the more violent of creatures.

Even during the overwhelming violence of World War II, in which some sixteen million persons were killed either directly or indirectly in seven years, the rate of killing does not approach that of Schaller's lions.

Although other wars have also taken similarly horrifying tolls, it is still true that, over the long run, including the relatively long periods when there were no major wars, the rate of human killing averages out to be considerably less than that among lions and all the other large animals studied to date.

None of this should be taken to mean that the level of violence among human beings is in any way excusable. The killing of a single human being is a deplorable act warranting every possible effort to prevent further violence. Still, there is little to be gained by focusing on the acts of a very small minority of human beings—and using them to suggest that our entire species is violent or sick.

There is ample ground for wondering why the American murder rate is twenty times that of Britain but there is no evidence to conclude that human beings as a whole, or even Americans alone, are unusually prone to violence when compared with other members of the animal kingdom.

It is no surprise that people should consider the human race unusually violent toward itself. We have the ability to empathize strongly with victims of violence. The brutal slaying of a child, reported in great detail in a newspaper or reflected in the mother's sobbing on television, arouses in us strong feelings of revulsion toward the perpetrator. Through the mass media a single killing may become known to millions. Over a month's time it would not be unusual to learn in considerable detail of a dozen human murders in one's own city or state. Our thinking about these killings seldom includes a calculation of the millions of unharmed people among whom the murder victims lived.

When confronted with the fact of many individual killings it seems natural to wonder whether there is not some violent streak in man, whether we do not bear the "curse of Cain," about which people have long philosophized.

Over the past fifteen years the idea of man's murderousness has taken on the appearance of scientific credibility largely through Ardrey's *African Genesis,* which popularized an old theory evolved by Raymond Dart, a professor of anatomy at the University of the Witwatersrand in Johannesburg, South Africa. Dart had found a number of fossil skulls of ape men and baboons which he contended showed evidence of early man's propensity to bash in the skulls of baboons and his fellow early men. Dart held that man evolved from a "killer ape," whose murderous instincts remain deeply ingrained in us, despite a veneer of civility. For many of the thousands who read Ardrey's book and of the millions who heard about it, the theory remains a chilling but plausible explanation for the apparent brutality of modern man. The theme of man's innate depravity has been further dramatized in such works as William Golding's *Lord of the Flies* (1955). The idea even offers some escape from guilt: if we are born this way, we can't help it. The "killer ape" hypothesis, however, is an idea whose time has gone. In fact, most of the leading scientists who study the evolution of man say the theory was never widely accepted. Despite Ardrey's skillful use of his dramatist's talents to tell a compelling tale, few scientists have found enough hard evidence to support the notion. But the impact of Ardrey's book has been so pervasive and, in my view, so damaging, that it warrants detailed examination.

The blood-drenched story that Ardrey presents as inescapable scientific truth has its roots not in evidence and not in Africa, but in Ardrey's own emotions about Africa. Ardrey's visit to Africa, during which he conceived much of the book, was initially to cover Kenya's Mau Mau uprising in the early 1950s for *The Reporter* magazine (now defunct). His first exposure to the fossils, upon the ambiguous features of which much of his argument rests, was mixed with hearing tales from white settlers terrified by Mau Mau killings. The black freedom fighters had taken secret oaths in the dark of night and amid grisly rites of sacrifice. They had shed the European's cloak of civilization, the settlers believed, and reverted to the level of animals crazed by the taste of blood. It was much the same exaggerated story from which Robert Ruark spun his allegedly based-on-fact novels.

"I sampled in the terror-brightened streets of Nairobi the primal dreads of a primal continent," Ardrey wrote in *African Genesis*. "I learned to fear for my life in a thousand ways, and in a thousand moments to yearn for the mortal security of civilization. . . . Africa scared me. If this continent had indeed been the cradle of humankind, and I had been the first man then I should have been born in fear."

The book goes on to make the case that *Homo ardreyensis* could have survived only by using the deadliest weapons he could find or make to defend himself and to slay his murderous brothers lurking in the dark. The ape men who could make the best weapons or use them most skillfully survived; those who could not, perished, the victims of the weapons. Ardrey argued that the instinct to kill survives in its purest form in adolescent street gangs where vibrant youth, still unfettered by the artificial restraints of culture, remained as dedicated to the switchblade knife as its ancestors were to the bone club or the flint-edged dagger. Ardrey contended that young thugs such as those of *West Side Story* ("No other work of art . . . offers such a vivid portrayal of the natural man") are mentally healthier than civilized adults because they lived in harmony with their animal instincts.

"Man is a predator whose natural instinct is to kill with a weapon," Ardrey declared. On the larger scale, this instinct reveals itself in mankind's giant organizations and machines for making war. In fact, Ardrey argued, the histories of man and lethal weapon have been so intertwined for so long that man has become utterly dependent on weapons and cannot live without them.

Most of the philosophical and sociological premises and consequences of Ardrey's beliefs were rather ably refuted in a little 1968 volume titled *Man and Aggression,* edited by Ashley Montagu, the Princeton University anthropologist. It is a collection of essays, most of which appeared earlier in general periodicals. Yet, strangely, the book has no contribution by an expert on the fossils that first sparked Ardrey's imagination. Much of Ardrey's case rests on the finding of some fossil skulls and jaws of *Australopithecus* (a species of ape man then considered to be a human ancestor) that were broken in what appeared to have been assaults by fellow australopithecines. Murder, in other words.

In 1953, two years before Ardrey began his study in the subject, Dart published a paper entitled "The Predatory Transition from Ape to Man." Dart argued that a certain evolving line of apes diverged from the typically herbivorous habits of its predecessors and acquired a taste for meat which it hunted as a predator. As Dart saw it, the apes evolved the upright posture, which made possible the dextrous hand, which made possible the fashioning and use of weapons, which made possible the more efficient killing of prey. Man emerged from apedom because he was a hunter, Dart said. The demands of hunting made opportunities for increasingly intelligent creatures to prosper and become still better hunters. That much accords reasonably well with today's views, but Dart went further. He argued that it was but a short step from killing animals for food to murdering one's own kind. As the near-men came to rely more and more on their improving weapons, Dart implied, they evolved to become more at home with them and came rather to like the killing business. The ones who had the best weapons—also the ones with the sharpest minds—and the strongest stomachs prospered and had no qualms about murdering their less violent brothers who, they may have felt, were holding back progress.

In his paper Dart wrote, rather verbosely, "The blood-bespattered, slaughter-gutted archives of human history from the earliest Egyptian and Sumerian records to the most recent atrocities of the Second World War accord with early universal cannibalism, with animal and human sacrificial practices or their substitutes in formalized religions and with world-wide scalping, head hunting, body-mutilating and necrophiliac practices of mankind in proclaiming this common blood-lust differentiator—this predacious habit, this mark of Cain that separates man dietetically from his anthropoidal relatives and allies him rather with the deadliest of Carnivora."

That paper could hardly have had greater impact than it had when read or pondered in the heart of "primal" Africa during the time of the Mau Mau rebellion—in the very place where man was born. Building on Dart's views, Ardrey contended that near-man's acquisition of weapons (necessary in the absence of sharp teeth or claws) became the all-absorbing focus of his existence. In Ardrey's view, the weapon became and still remains man's "most significant cultural endowment." In part this conclusion was based on Ardrey's view that most of the cultural artifacts found with the remains of early men were not merely tools, as most archaeologists call them, but weapons. Ardrey considered "tools" a euphemism. That view does not hold up. The choppers, scrapers, burins, awls, hammerstones, anvils, handaxes, and flakes of various sorts that were found would have made poor weapons with which to kill anything. Most authorities believe they served a variety of household utility functions. Of course, no one doubts that an early hominid could have taken a chopper, originally fashioned for another purpose, and bashed his brother on the head. It could have been as dangerous as a baseball bat or a tire iron. Undoubtedly early man had weapons with which to kill his prey. He probably used wooden clubs (which have disappeared because they do not fossilize well) or threw rocks (which do not deteriorate but look like any other rocks). One probable method of killing, deduced from the circumstances in which butchered and fossilized animals have been found, was to drive the prey into muddy ground, which would hold the animal almost immobile until it could be clubbed or stoned to death. Such a killing method relied more on cunning and planning than on the invention of a superior material weapon. Long before the first evidence of spears or arrows, man was undoubtedly an accomplished hunter. He relied not on fancy weapons, but on his brain to make snares, or to drive animals over cliffs or into pits, or to surprise them while they slept. It also seems likely that early man would have tried to steal the kills of lions, hyenas, and other predators. Various animal predators steal the kills of one another quite often, and no doubt our ancestors saw how easy it was. Modern people on foot can easily scare lions off kills with a few shouts and waves of the arms.

Among the fossil remains that Dart offered and which Ardrey found so compelling were a number of baboon skulls found along with the bones of *Australopithecus* and bearing the damage of a blunt instrument. The clear suggestion was that the ape men bashed the baboons' skulls with a weapon and ate their flesh. Ardrey's book ominously declared from this evidence, "The use of weapons had preceded man." Still, hunting is not murder. The fossil clincher for Ardrey was the jawbone of a twelve-year-old *Australopithecus* which had been broken before its owner died. A dent in the chin had knocked out some front teeth and the bone was broken on either side. Ardrey said that because the break had not knitted, the blow on the chin was fatal. Although it seems reasonable to conclude that the creature could have died shortly before or after the blow on the chin, Ardrey does not discuss the possibility, even the probability, that death came from damage to some more vital organ than the chin. There is no evidence of how the jaw damage might have been inflicted. A fall is one possibility that would involve no foul play. "What if a weapon had done this deed?" Ardrey wrote breathlessly. "What if I held in my hands the evidence of antique murder committed with a deadly weapon a quarter of a million years before the time of man?"

Ardrey also cited seven other ape man skulls that had been fractured, but even he cautioned, "Not all the specimens demonstrate conclusively death by purposeful violence." He singled out only three of the seven—one that had been bashed in a way resembling the damaged baboon skulls, one in which a two-inch rock had been rammed through the skull, coming to rest in the brain case, and one which bore two small punctures that could have been inflicted with a sharp stick. Ardrey did not suggest the nature of the weapon that could have done the damage to the third skull and discounted the possibility, widely accepted by many authorities, that the two holes were tooth marks from a predator. In fact, it has been shown that the two holes are at exactly the right distance apart to have been made by the long canine teeth of a leopard dragging its prey by the head, as leopards often do. Ardrey said one did not need more evidence or more expertise to determine what had happened.

Of the broken jaw of the twelve-year-old mentioned above, he insisted, "One needed nothing but the lay common sense of a juryman to return a verdict that at some terrible moment in most ancient times, murder had been done." That is the sum total of the direct evidence for the notion that murder and an affinity for lethal weapons was an instinctive, species-wide trait among man's ancestors and, hence, among us. That the bone damage might have resulted from a fall or from some other accident (ceilings had sometimes collapsed in the South African caves in which the fossils were found) was not considered a likely possibility. The balance of Ardrey's 357-page book is taken up with indirect suggestive evidence and descriptions of territorial and aggressive behavior among animals far removed from man's line of evolution. Curiously enough, Ardrey discounts behavioral studies of man's closest living relatives, the chimpanzees, which are remarkably amicable among themselves, even though, as Jane Goodall has shown, they do hunt and kill other animals for food on rare occasions.

African Genesis was published just as the wealth of discoveries by the Kenyan anthropologists Louis and Mary Leakey at Olduvai Gorge in Tanzania began to come in and long before other anthropologists began their searches in other parts of Kenya and Ethiopia, which have yielded many well-preserved remains of early man. The vast bulk of today's knowledge about man's evolution was discovered and studied since Ardrey's book came out. In fact, *African Genesis* turned so many people's minds to Africa as man's place of origin that it should at least be given credit for stimulating much of the financial support that has made possible the later work. How has the "killer ape" theory held up in light of this newer and more abundant evidence? Very simply, it has not. The additional fossil evidence that might have been expected, had Ardrey and Dart been right, has not been found among the hundreds of additional hominid fossils that have been found.

Today the vast majority of experts familiar with the fossil evidence agree that while man's ancestors appear to have been hunters who killed animals and consumed quantities of meat, there is no suggestion they were driven by a blood lust any more than any other predator. More important, there is certainly no hint that they killed each other more than does any animal species known today. As mentioned earlier, the evidence from the other species shows that man is among the least likely to kill his own kind.

"The killer ape theory should have been abandoned by everybody long ago because there's really no evidence any longer to substantiate the view," said Richard Leakey, the son of Louis and Mary Leakey. The younger Leakey's own discoveries in the fabulously rich area east of Lake Turkana (formerly Lake Rudolf) in Kenya have provided a large share of the known fossil evidence of early man. "The evidence for a predatory early hominid is perfectly valid," Leakey added, "but a predator ape is not a killer in the sense that the 'killer ape' was introduced and is popularly conceived of. I think it is very likely that nobody would have heard of the killer ape concept if it hadn't been for Robert Ardrey, but in the same breath I will say it is certain that a majority of the people now interested in supporting early man research would never have heard of early man if it hadn't been for Ardrey's book. I think Robert Ardrey has done a tremendous amount to bring in the interests of people around the world and if the price was the killer ape hypothesis, then I think it's cheap at the price."

Mary Leakey and F. Clark Howell, an American authority on human evolution, who has discovered many important fossils in Ethiopia, say much the same thing.

Phillip Tobias, a student of Dart and his successor at the University of the Witwatersrand, said, "I don't go along with this killer ape hypothesis myself. I don't think there's any good evidence to regard these creatures as having been unduly murderous in instinct." Tobias, an authority on both the South African fossils and those of Olduvai Gorge, believes the teeth of early human beings and the food remains in their camps suggest they were omnivorous, eating not only meat but a fair amount of vegetable matter as well. "There's quite a lot of evidence," he says, "of muddled thinking in the claims which attribute these murderous, bloodthirsty, slavering, meat-hungry instincts to the early hominids." Tobias feels it is fallacious to suppose that the behavior patterns of man today could invariably be guided by the same forces that moved australopithecines. "I believe," he says, "there is a certain amount of mixing or even confusing of levels of organization to jump straight from *Australopithecus* to *West Side Story.*" Tobias believes there are indications that the evolutionary steps that took place between man's ancestors and modern man included quantum jumps in mental ability that, among other things, conferred an ethical sense quite unlike anything in lower animals.

There is another line of evidence against the Dart-Ardrey contentions. Even if one accepts that the australopithecines were as murderous as the pair said, there is growing evidence that these ape men were not our ancestors. When Ardrey wrote, virtually everyone in the field considered *Australopithecus* a likely human ancestor. Since then, new fossils have come to light in Tanzania, Kenya, and Ethiopia that suggest man had already come into being before any of the *Australopithecus* individuals known from fossils were born. Some authorities now believe that both true man (*Homo sapiens*) and the *Australopithecus* ape man are descended from a common ancestor that remains unknown to science. The oldest known members of our own genus, *Homo,* lived more than a million years before the oldest known members of the genus *Australopithecus.*

The idea that man is innately violent has also been widely promoted in the writings of Konrad Lorenz, the Austrian zoologist who won a Nobel Prize in physiology and medicine in 1973 for his work on ethology, a kind of animal behavior research that dwells on social behavior within a species. In a nutshell, Lorenz's thesis, best known from his book *On Aggression* (1966), was that aggression among the animals serves to establish leadership for the defense of the group and to distribute individuals equitably over the habitat. However, because actual bloodletting or killing would serve to weaken or reduce the species, many animals have developed rituals that take the place of violence. After a session of ritual sparring, one animal accepts defeat and make some gesture of capitulation, sometimes deliberately exposing its most vulnerable anatomic region, and the victor, recognizing the capitulation, magnanimously refrains from further attack.

Lorenz contends that human beings have failed to develop much of an aggression ritual and that, in any case, our advanced weapons now make it easier to kill over distances that rule out face-to-face communication. With the development of the bow and arrow, perhaps even the accurately hurled stone, it became possible for one person to kill another without ever having to look him in the eye. The gun, the bomber plane, and the intercontinental rocket have moved combatants still farther apart. Hence, according to conventional Lorenzian thinking, mankind should be getting worse and worse in levels of violence. Many people, including Lorenz's large popular following, see little that would move them to doubt this.

At a major symposium on "The Natural History of Aggression" held at the prestigious British Museum, Natural History, the views of Ardrey and Lorenz received virtually unanimous scientific acclaim in 1963. The scientific community has moved away from this position in recent years, but it has been only slowly relinquished in the popular view.

During that Natural History Museum symposium, scientists spoke of "the unimagined strata of malignity in the human heart" and no less a figure than Sir Julian Huxley, the eminent biologist, rose to suggest there was an embarrassing contrast between man's behavior and the kindliness with which lions behave toward one another. George Schaller's findings on lions, of course, had not yet been published.

The British magazine *Nature,* perhaps the leading general scientific journal in the English language, carried a summary of the meeting that rather well presaged the opinions that would come to be popularly accepted during the years that followed:

Certain tentative generalizations can be made. The irrefutable and terrifying history of overt aggression appears to be essentially human; animals display aggressive attitudes which may have survival value, but under natural conditions they do not fight to the death with members of their own species; aggression is ritualized so that little damage is done. Man's beastliness is not of the beast; to the anthropologist and the historian, human, overt aggression may seem normal but against the background of the animal kingdom, from a point of view which cannot be avoided by the biologist, it appears pathological.

According to Edward O. Wilson, the Harvard zoologist, Lorenz's writings pointed the wrong way: "The evidence of murder and cannibalism in mammals and other vertebrates has now accumulated to the point that we must completely reverse the conclusion advanced by Konrad Lorenz in his book *On Aggression,* which subsequent popular writers have proceeded to consolidate as part of the conventional wisdom."

I would argue that what many of the leading scientists thought just a few years ago and what a leading journal relayed with little hint of doubt, is now plainly wrong. We are not descended from killer ancestors; we are not the bloodiest of creatures on the earth; we do not bear the mark of Cain. We *are* descended from apelike creatures that hunted animals for food; we *are* among the most pacific of the larger animal species; and, if we bear any "mark" at all, it is a capacity for empathy, for kindness, for decency, for what may simply be called humane behavior. There is plenty of evidence that our species advanced not through conflict but through co-operation.

In one sense it may be comforting to believe that we are a comparatively violent species by instinct. If that's the way we are, then there is not much we can do about it, and the burden of the violence of which we are truly guilty can rest squarely on a genetic inheritance over which we had no control.

On the other hand, to accept that we are not violent by instinct is a much more challenging proposition. The burden then falls on us to be responsible for the violence that we *do* commit. Our violence is of our own making and, therefore, for us to unmake.

Man is not some miserably depraved species. He is another of the animals, considerably kinder to his own species than most are to theirs and vastly more capable of the empathy that can and should move him to be still kinder.

13

The Cult of the Wild

The campaign to conserve wildlife for the enjoyment of future generations is not a modern phenomenon. In ancient Egypt about 4,500 years ago, for example, one King Sahure enforced the protection of certain wild animals that, it was feared, might otherwise be decimated. In ancient Persia, now Iran, about 2,500 years ago there were huge, walled game parks containing a variety of wild animals living free. Some 2,000 years ago in Syria the royal government took steps to protect lions, which then could be found there, and a remnant herd of Asiatic elephants. In what may be the most successful sustained formal conservation effort to date, the Syrian elephant herd survived for six centuries.

Wildlife conservation programs were also instituted in many parts of Europe for the last thousand years or so, particularly in England and Germany which, to this day, are the European countries most active in modern wildlife appreciation.

Before saying more about the history of wild animal conservation, one crucial fact, common to all the programs just mentioned, needs to be added. They had, at least in their early days, nothing to do with ecology or a love of nature. The goal of these efforts was to ensure the continuing availability of animals to be killed by sport hunters. Like it or not, it was the people who got a kick out of killing animals who laid the historical basis for wildlife conservation. And it is their attitudes—sometimes sensible, sometimes not—that have largely guided conservation practices. Even to this day, anyone who peruses the lists of boards of directors for the major conservation groups and inquires into personal backgrounds will find that one of the largest single special-interest groups represented is that of sport hunters or, in many cases, former sport hunters.

It is only comparatively recently that wildlife conservation for purely aesthetic or ecological reasons has become a real force, rallying large numbers of followers. There is, however, an interesting difference between the leaders of the conservation movement and their new followers. The leaders are, by and large, people who have had considerable contact with wild animals and the wilderness, while the followers are people with little or no personal experience with the wild. The two groups have different attitudes on many issues, the most obvious of which is hunting. Conservation movement followers are generally strongly antihunter; they regard sport hunting, and sometimes even subsistence hunting, as abhorrent and something to be done away with. Many conservation movement leaders, on the other hand, are not opposed to sport hunting and, in fact, regard it as a perfectly reasonable pursuit so long, of course, as it is done legally. The leaders, naturally enough, do not go out of their way to make their prohunting views known to their followers. The rank and file of the conservation movement, drawn largely from cityfolk, are sensitive people whose continuing support of the movement often depends on their preserving their idealized views of how people should relate to animals. This idealized view is catered to most obviously in the magazines, films, calendars, picture books, and other products put out by the conservation groups. In fact, it could be argued that these products present such a carefully sanitized view of wildlife and wilderness that many supporters of conservation continue to have a rather unrealistic idea of what "nature" is. The sharply focused, dramatically lighted photographs almost never show you a tick-studded antelope or a scarred and mangy lion. How often do they portray wolves gobbling up some rotting carcass, maggots and all? The "proud" and "magnificent" bald eagle is usually shown in a pose befitting a national emblem and not in the act of scavenging bloated and beached fish. Lions or elephants are never shown killing people, which both do from time to time. Quite often, whenever anything unappealing is depicted, it is something man-made—smog over a big city, gunk and junk floating down a river outside a factory, a bulldozer pushing down virgin forests. This unrealistic attitude in which man and nature are implied to be fundamentally inimical is not new, of course. It can be seen in a painting entitled "Le Paradis Terrestre" that was painted nearly four centuries ago by the Flemish artist Jan Brueghel (1568–1625), or in any of a dozen "Peaceable Kingdoms" with the scenes of lambs and lions reposing together. This is how it supposedly was in the Garden of Eden before man ruined things. The untouched wilderness is a paradise, this theory goes, but put man in the picture and it is suddenly unnatural, unappealing, evil.

The survival of the sport hunters' conception of wild animals in modern conservation can be seen in the name by which everyone, even aesthetic conservationists, call wild animals: game. In Africa the wilderness preserves are popularly known as "game parks." This is not so much the case in the United States because the national parks are more often regions protected for their spectacular scenery rather than for their wildlife. The concept of "game" as it has evolved over the years is worth exploring in any attempt to understand the history of wildlife conservation. Game, in its original context, is whatever sport hunters believed worthy of their bullets. Traditionally, among hunters any animal species that is not deemed game is called "vermin." Game animals are good; vermin animals are bad. There are rules of etiquette that govern the killing of game; none that limit the killing of vermin. There is also a rule that only a privileged class of people may shoot game. Anyone who does not belong to this class is called a "poacher." Until the last century or so, the only persons permitted by the game laws of many countries to kill game were those of royal or noble birth. Just as the European kings ruled by divine right and believed their realms to be their personal property, so the hunting of game, often called "royal game," was their right alone. Royalty even asserted that the individual game animals belonged to them personally and the laws protecting these beliefs were enforced by game wardens. Any poor English farmer of those days who drove off the stag that raided his crops or, God forbid, shot or trapped a boar to feed his family was automatically a "poacher." If caught, punishment was severe, often fatal. What the commoner was engaging in was, of course, hunting, plain ordinary hunting to sustain one's livelihood as human beings have done for at least two or three million years. We are a hunting species and only in the last fraction of 1 per cent of our history have we learned to get our meat by killing domesticated animals rather than wild ones.

This upper-class distinction between hunting and poaching continues today, particularly in Africa where most people still make their living from the land, a significant minority being chiefly hunters. The game is no longer "royal." It is "national" or even "mankind's heritage." But killing it to sustain one's family is still poaching. Even among politically liberal Americans, who like to think they have some sympathy for the problems of poor people, it is quite fashionable to damn the poachers in Africa who shoot or spear or trap wild animals. Recently one of the largest conservation groups in East Africa, supported chiefly by American donations, put out posters soliciting contributions with the slogan, "Help stop the senseless slaughter by poachers of East Africa's wildlife." The hunting of food or of animal products that can be exchanged for money by impoverished Africans is hardly "senseless." It is, in fact, quite sensible, as it is the only source of livelihood open to many and, until the advent of the European with his concept of royal or protected game, wildlife was considered unownable by any individual or class and, therefore, as free for the taking as air or water.

The classifications of game and vermin have not been im-

mutable. At one time in England game amounted to little
more than deer and boar. Predators such as wolves, wildcats,
and foxes were vermin and their extermination was a job for
professional vermin hunters who were deemed to be perform-
ing a service. The predators were particularly disliked because
they competed with sport hunters for the game. By the seventeenth century wild deer and boar were becoming rare in England (as were the predators) and hunters turned increasingly
to smaller game animals such as hares, rabbits, and certain
birds. These didn't provide quite the challenge of the chase
the deer did and so the classification of "game" was enlarged.
Foxes, once pests to be exterminated, became game and steps
were taken to ensure a good supply for hunters. Commoners
who once could shoot a fox for stealing chickens now risked
punishment for protecting their domestic animals. While the
poor man had to sacrifice to help conserve foxes, the rich
turned the fox hunt into one of the most ritualized forms of
"sporting" recreation the Western world has seen, what with
gaily adorned horses, carefully groomed hounds, *de rigueur*
costumes for the people, special trumpet flourishes to mark
key events in the hunt, and, of course, the mandatory cry of
"Tally ho!" Often, to make sure the great social event went off
as planned, the foxes were captured beforehand or obtained
from special fox parks and released near to where the hunt
would begin.

As foxes began to decline in number, emphasis shifted to
bird hunting in which, owing to the comparatively small size of
most birds and still less of a chase being possible, success was
measured by the quantity of animals killed, not necessarily
the quality of the way in which they were brought down. This
obviously took a large toll on the wild pheasant and partridge
populations and, in the late eighteenth century, pheasant and

partridge farms sprang up to supply the numbers needed to satisfy bird shooters. The Prince of Wales (later to become King Edward VII, who died in 1910) and nine companions, shooting in Hungary, were reported to have bagged an almost unbelievable twenty thousand partridges in ten days.

The formal distinction between game and vermin, heroes and villains, has persisted into the present, even as the emphasis of wildlife conservation gradually shifted from supporting sport hunting to preserving for ecological and aesthetic reasons. The distinction was widely observed in East Africa where the British tradition held sway through the colonial period. Remnants of it remained strong even after Kenya, Tanzania, Uganda, and other countries with abundant wildlife gained their independence in the 1960s. In the original 1904 game laws of Kenya the "vermin" category included lions, leopards, cheetahs, hyenas, and baboons. The predators threatened the game (meaning mostly the antelopes) and the baboon and hyena, neither then known as predators, were simply thrown in, presumably because they were thought repulsive.

The annual reports of the colonial game departments in Kenya and Uganda make clear that one of the departments' prime functions was to protect the animals wanted by sport hunters. In 1924 Captain A. Ritchie became the head of the Kenya Game Department and shaped its philosophies until his retirement in 1948. He actually hired vermin hunters to go out and shoot offending animals. One of his vermin-control officers, an annual report boasts, shot eighty lions and ten leopards. Another, reported Ritchie, as we have seen in Chapter 7, on baboons, "worked well and consistently," poisoning thousands of animals such as leopards, hyenas, baboons, and bush pigs. In its eagerness to eliminate pestiferous

animals, the Game Department sometimes believed it necessary to "control" certain game animals that were deemed tem-
porary pests. Buffalo, for example, sometimes bothered people and had to be shot. However, Ritchie's efforts to recruit
men to go out and get rid of them en masse failed because it
was decided to be too dangerous an operation. Buffalo,
Ritchie wrote in desperation, "would get short shrift if any
feasible means could be found to destroy them." Rhinoceroses
were not much valued in the living state; their sawed-off
horns, however, brought fancy prices from traders who ex-
ported them to the Orient, where powdered rhino horn was
and still is considered an aphrodisiac. In 1943 alone, the
Game Department "controlled" over a thousand rhinos. Even
elephants sometimes misbehaved and had to be controlled.
"Experienced elephant hunters," one report said, "were let
loose in the area and about 200 elephants were shot. . . .
Lest sportsmen bewail the slaughter it must be pointed out
that nothing except trash was shot." "Trash" presumably
meant elephants whose tusks were of less than desirable size
—in other words, younger elephants.

In Uganda in the 1920s elephant control took on the nature
of a paramilitary operation. In fairness, it must be stated that
Uganda's once-widespread elephant population did raid
farmers' crops and occasionally menace villagers. In 1924,
when the Uganda Game Department was established, elephants occupied nearly 75 per cent of the land area. As the
human population increased, clearing more land for farming,
the elephants' ranges were compressed, diminishing the total
area of the country containing elephants but increasing the
concentration in any one area. This led inevitably to conflict
between Ugandan farmers and elephants and to the impression that elephant numbers were increasing when, in fact,

they were almost certainly declining. Today elephants occupy but 13 per cent of Uganda. In 1925 the first head of Uganda's Game Department, the famed Captain Pitman (like most early wardens in East Africa he was a military man and a hunter), wrote in his annual report as if he were battling an insurgency, "There is still plenty of room in Uganda for both the [human] population and the elephants, but a comparatively large number of elephants must be destroyed each year to prevent them overrunning the country." Accordingly, Pitman drew up a map of Uganda, dividing it into elephant zones and people zones. "In other words," Pitman wrote, "a system of defensive fronts has been instituted to ward off the encroachment of elephants on the more thickly inhabited areas." Pitman allowed himself to get bogged down in a land war with the elephants and it dragged on for years with no clear-cut victory. "The whole secret of success is mobility," Pitman wrote. "The elephants are becoming wise to control tactics, and adapt themselves accordingly. They move swiftly and silently raid the shambas [African farms] in the dark, and are usually far away before the first streak of dawn." In subsequent annual reports Pitman concedes that elephant "control" may be leading to increasingly uncontrollable elephants. "The ranger," he wrote of one problem area, "considers the possibility that the Singo elephants have had enough shooting for the present. Excessive hunting of elephants is more conducive to shamba damage than no hunting at all."

With the advantage of hindsight it is now clear that the white man's stewardship of wildlife in East Africa did little to protect people from the genuine depredations of wild animals and much to reduce the numbers of many species.

Just because an animal was not classed as game did not, of

course, mean that hunters did not hunt the species. It only meant that a species was so common that hunting it was not considered very sporting and not worth regulating. Lion hunting, for example, had been a tradition for centuries. Before the species became classified as a game animal in East Africa, anyone was permitted to go out and bag as many as he wanted any time. In 1910, according to Guggisberg, the early lion expert, an American using dogs shot sixty lions on a single safari. Gradually, however, wildlife authorities recognized that there was growing interest in lion hunting and that maybe something should be done to ensure that there were lions left to hunt before the unregulated killing of the species eliminated them as an attraction for hunters. In Kenya in the 1920s Ritchie wrote, "I am of the opinion that it will be necessary in the near future to limit the number of lions that may be killed on a full license. Otherwise this animal will soon cease to exist in some areas where he does no harm but—to put the matter on the most sordid basis—helps to sell revenue."

As the money-making possibilities of East African wildlife came to be recognized, more species were taken out of the "vermin," or unclassified, status and made "game." This meant that they could only be legally hunted on licenses sold specifically for the species to persons who paid a set fee. As the numbers of any species declined and the fee for that species increased, the lure to the sport hunter grew. Until quite recently, when hunters as a group became more ecologically minded, the sportsman's greatest pride was in bagging the rare animal. The rarer and harder to find, the greater the challenge to the hunter and the bigger the ego once the trophy had been pinned to the wall. As a side effect, the reclassification in some bureaucrat's office of an animal as "game" turned many

Africans, who had been hunting far out in the bush as usual, into poachers. Overnight, and with scant attention to its impact on Africans, the changing of the game laws turned law-abiding citizens into criminals.

Another animal that illustrates the way in which conservation laws have been shaped by unfounded human attitudes toward a species is the crocodile. As we have seen, crocodiles have been universally hated. Until quite recently it was the unspoken duty of every good game warden to shoot any crocodile on sight. Extermination was the implicit long-range goal. When Pitman rewrote some of the Uganda game laws to ban the shooting of wild animals from vehicles (this was considered unsporting), he made an exception if the target was a crocodile. Today crocodiles have become part of what is often called Africa's "priceless wildlife heritage"; all killing of crocodiles is banned in many parts of East Africa.

Although today most of those who deal professionally with East African wildlife have accepted the ecological reasons for conserving wild species and largely abandoned the old distinctions between "good" and "bad" animals, there remain a few prescientific notions. For example, the British warden of one national park in Kenya still doesn't like baboons. The warden who, like many, lives in the park, regularly puts poisoned bait around his house to get rid of the vermin. It was not too many years ago when another warden successfully poisoned the hyenas in Nairobi National Park.

By the 1950s the interest in wildlife conservation in East Africa began to change. The influence of the old-time hunters and game wardens declined and a more rational appreciation of aesthetics and ecological principles grew. In some cases the old-timers simply died out or moved away. In other cases they changed their point of view, putting away their early interest in hunting and exchanging their advocacy of conservation to benefit hunters for conservation to benefit the habitat and/or tourism. Of crucial significance was the post-World War II economic boom in Europe and the United States which gave unprecedented numbers of people the money to visit Africa. The money-making possibilities of wildlife shifted from the value of ivory and horn and the sale of hunting licenses to tourism. Suddenly East Africa was receiving not the politicians and tycoons who wanted to hunt but ordinary people, albeit of some means, who wanted nothing more than to look at the animals and take their pictures. Game or vermin, it made little difference to the new and better-educated wildlife constituency. Americans who grew up in a land in which earlier Americans had exterminated most large animals began to regard African wildlife as not just African but their own as well. They transferred African animals from the province of Africans and of big game hunters into the "priceless heritage of mankind" everywhere. The conservation leadership that had already grown up to advance the cause of hunters shifted its character to take advantage of the new sources of support, financial and otherwise, that lay in the broader constituency.

In the 1960s the fledgling East African nations, no longer British colonies, had to build independent economic foundations, and tourism was an obvious first step. The British had established a small number of national parks, but the African governments quickly created more and enlarged some of the old parks. Contrary to the fear of some colonialists that the Africans would neglect the wildlife, exactly the opposite has happened. In Tanzania, for example, the British left only one national park, Serengeti. In fact, the last British governor-general cut off a section of the park just two years before independence. Today, Tanzania has ten national parks comprising 3 per cent of the country's area, and President Julius Nyerere has enlarged Serengeti.

"Wretchedly poor as it is," Bernhard Grzimek, the famed German naturalist and early advocate of protecting the Serengeti region, wrote in his book *Among Animals of Africa* (1971), "Tanzania now devotes a larger proportion of its national income to the conservation of natural resources than the United States. Developments are taking a similar course in other countries of East Africa. Young African nations are shaming the older countries of the white man, in which many species have become permanently extinct. This is especially humiliating to a citizen of the German Federal Republic, whose politicians have yet to establish a single national park or wildlife sanctuary and devote virtually no funds to the conservation of natural resources."

Much as Americans may take pride in the leadership they have given to world-wide efforts to conserve animals elsewhere, the United States Government effort to conserve its own wildlife for purely ecological or aesthetic reasons is also a comparatively recent phenomenon. Hunters, of course, have long been interested in state or federal efforts to assure them something to hunt, and their license fees and other special taxes have paid a large share of the costs of maintaining state fish and wildlife departments for many years. There *was* a brief flurry of conservation activity for nonhunting reasons around the turn of the century when people realized the bison might become extinct. The U. S. Army's nineteenth-century campaign to exterminate the bison to deny food to the Indians was a major factor in reducing North America's once dominant herbivore from perhaps twenty-five million in the 1840s to under six hundred by 1889. Individuals concerned about the bison's fate eventually won a political battle and the animal was given protected status. It has, of course, increased in numbers and by now there are several thousand.

The largest remnant of the original bison in 1889 was in the Yellowstone National Park, created in 1872 as the first national park in the United States. The notion that the creation of Yellowstone more than a century ago was an early manifestation of government interest in wildlife conservation is, despite frequent boasts to that effect, wrong. The handful of real conservationists who wanted to set aside the Yellowstone area were never able to get Congress to share their enthusiasm for the wilderness. Congress, according to some accounts, acted only when lobbyists convinced the legislators that tourism (to see the geysers) would earn more money for the region than would the sale of mineral rights. Even the conservationists who worked for the creation of this and other early national parks were not as much interested in preserving wild animals as they were in protecting areas of great scenic beauty that they felt would be appreciated by city folk grown weary of the concrete jungles. Frederick Law Olmsted, one of the country's early leaders of the national park movement, clearly thought of such parks as serving the same function as did his Central Park in New York City—a place to refresh the urbanized human spirit.

Another early step in conservation for ecological or aesthetic reasons was the passage in 1900 of the Lacey Act which in effect protected egrets and herons whose delicate plumes doomed millions of the birds to the millinery trade. The act made it a federal offense to ship across state lines birds whose hunting was limited by state law. Through vigorous enforcement, the act has been responsible for the present relative abundance of these birds, particularly in the Florida Everglades.

After these early starts, however, the federal government lost its interest in wildlife conservation and virtually nothing was done for purely aesthetic or ecological reasons until about a decade ago when Congress passed the Endangered Species act of 1966. The act, however, applied mainly to species that were already so reduced in numbers that any ecological impact from their disappearance had already been suffered. The law's chief purpose, then, was aesthetic in satisfying the desires of the new wildlife constituency that wants simply to know of an animal species that it is there. However tenuous the species' hold on survival, the law allows us to take a measure of satisfaction in knowing that we are not yet guilty of wiping it out.

However laudable the 1966 act and its motivating sentiments may have been—and they are laudable—it has given rise among people to a curious new form of wildlife appreciation that may be called the "cult of the endangered species." Members of this cult make so much of endangered species that popular interest in them soars far above that in nonendangered species. Take, for example, the ivory-billed woodpecker, a species that is so endangered it may already be extinct. Thousands of wildlife enthusiasts would give their eye teeth to be able to spot one and thousands go to great efforts to do so in such places as the Big Thicket National Biological Preserve of East Texas where some say the species may still exist. As it happens, there is another species of woodpecker that doesn't look all that different, called the pileated woodpecker. It is not endangered, but far fewer people have heard of it and even fewer are eager to go out in the woods and appreciate one of these. The reason for this behavior, of course, is exactly the same that once motivated hunters to seek out the rare animals for their trophy rooms.

Perhaps the most outlandish expression of this new cult is the interest in saving the Devil's Hole pupfish, a species of quite unremarkable inch-long fish that lives in but one small pool in a Nevada cave. There never were any other places in which this fish lived; it is a local variant of pupfish that evolved into a distinct species because it was isolated from all other pupfish. There are thousands of such local variants of many kinds of animals all over the world. The Devil's Hole pupfish has no ecological significance beyond its own tiny pool where only about two hundred live. And yet there has been a substantial national battle to protect this species. Its chief threat is a nearby rancher who would like to pump more water out of his own well to irrigate his land. This, however, would reduce the water level in the pupfishes' pool, depriving them of much of their food. The battle has been taken to the United States Supreme Court and has been counted as meriting an impassioned editorial in the New York *Times*. In 1976 the Supreme Court ruled in favor of the pupfish. So much attention has come to these tiny beasts that they are far better known by wildlife enthusiasts and scientists than are many of the more common fish in American waters that are, by their very commonness, vastly more significant factors in their much larger ecosystems.

From an interest in conserving wild animals simply to shoot them later on, the interest in game saving has metamorphosed over the centuries to the point where there is a nearly total opposition to any thought of killing wild animals for any reason (witness the rise of the "animal liberation" movement, in which it is argued that people have no right to kill any animal). Modern conservation sentiment has also brought us to the point where the most prized species in the eyes of many are those that are the most endangered, or at least believed to be the most endangered. It matters little whether the animal has some intrinsic qualities that make it attractive or useful; rather it is the fact of endangerment that draws so much interest in it. This preoccupation with endangerment has led many conservationists to make almost unrelenting attacks on their own species, *Homo sapiens*. The endangered species is good, the conservationists' values suggest, while the endangering species is bad.

14

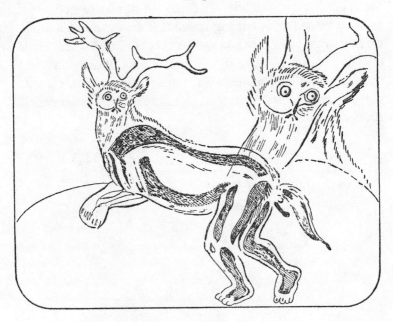

Toward a New Conservation Ethic

If we are to have wild animals living on the earth with us—truly wild and free-living beasts—then the time has come to put away childlike ideas about what animals are. There never was a Bambi or a Big Bad Wolf. Lions are not noble, brave, clean, thrifty, or reverent. The world would not be a better place without sharks or hyenas, or even without cockroaches or termites. Such notions about animals should disqualify a person from having any role in making the hard decisions that must be made in realistic and rational wildlife conservation programs.

This is no longer a world of natural luxury where we can afford every person's wildlife fantasies—letting all who wish to do so be hunters with gun or camera and also permit others who wish to do so—or must do so—be farmers or herdsmen or lumbermen. Just as the world has limited supplies of energy and food, it has limited supplies of the habitat that can support wild animals. And just as an expanding human population is pushing the demand for energy and food, so is it spreading to occupy more and more land that was originally wilderness.

Competition for land is by far the major threat to wildlife all over the world. Hunting, at least on land, is a relatively minor threat regardless of whether it is legal or illegal, for as long as a species' habitat is protected the survivors can almost always replace the killed individuals.

The environmental pressures on man and beast are now so great that it is no longer adequate, if it ever was, for advocates of wildlife conservation to justify their advocacy with childish or sentimental notions of what wild animals are all about. It is not enough to say that lions should be protected because they are such proud creatures or that whales should live because they are so intelligent. There is a valid place for aesthetic appreciation of wildlife and maybe even for a bit of fantasizing, but an argument for the protection of wildlife based on aesthetics or fantasy will not persuade the people who must be persuaded if genuine and lasting protection is to be secured. The argument for wildlife conservation must be clear-eyed and hard-headed. In many cases it must even be made economically attractive to those who are competing with the wild animals for their livelihoods.

Furthermore, in a world totally penetrated by man either through his own physical presence or through the chemicals and toxins man has spread over the globe, it is too late for the brand of conservation that advocates simply keeping hands off and letting nature take its course. We have already intervened too much in too many places—chiefly through the physical destruction of habitats—simply to walk away from the situation and assume that all will be right again. Except for a very few large wilderness areas, each comprising thousands of square miles that could theoretically be left alone, some form of wildlife management will be necessary to offset the disruptions already caused by human competition and the disruptions that will continue simply because most protected areas are likely to have artificial boundaries. Proper wildlife management is a complex business in which the action taken must be tailored to the needs of the species or habitat in question. This is why conservationists must know well the true behavior of an animal species. False and sentimental notions about animals can only cloud or retard intelligent conservation. Most professional wildlife conservationists know this already, but there are many examples of intelligent wildlife management programs put awry or killed entirely because they did not square with the popular sentiments of the nonprofessionals whose support is essential to success. The elephants of Tsavo National Park, in Kenya, are one example.

Another question demanding an answer is why we want animals conserved in the first place. If the reason is not a good one, it may not stand up to worthy alternative suggestions for the use of the land. The most useful arguments to be made in behalf of wildlife conservation are not that people shouldn't wear leopard-skin coats, or that poachers are greedy, or that hunters are on ego trips, or that men who club baby seals are cruel. Those make handy targets and it is fashionable to aim at them, but none of these exemplifies the major threat to wildlife. If you really want to protect as many wild animals as you can, you must argue that Tanzanian farmers must not try to enlarge their farms and grow more food. You must tell Brazilians not to try to raise their standard of living by extracting the mineral wealth of the Amazon basin. You must convince the peasants of India that they should be willing to sacrifice a year's income (the value of a bullock) in order that a Bengal tiger can have something to eat for a week.

What, then, is our reason for wanting to preserve wild animals that will convince these people of the Third World, where so much of the wildlife remains and where so much of it is threatened, that they should make sacrifices because *their* wild animals are part of *our* "precious natural heritage"? Or can we reasonably ask them to make the sacrifice without ourselves sharing the cost? Is it not time that we in the prosperous countries who are so interested in conservation find some effective way of sharing the economic burden of wildlife conservation? It isn't enough just buying subscriptions to wildlife magazines or donating to conservation groups. These things raise some money and do some good but not nearly enough to do the job. Wholly new ways must be found to pay for wildlife and even in some cases to make wildlife pay for itself. If we in the rich lands aren't prepared to put up, then we should shut up.

Let us examine the issues in more detail. In East Africa, the last great refuge of the kind of large mammals that once roamed much of the earth and that so fascinate Westerners, the human population is growing at rates exceeding 3 per cent a year. This is more than three times as fast as the United

States is growing. The populations of Kenya, Tanzania, and Uganda will double in the next twenty to twenty-two years and so will the need for food. Food means land and that means more habitat taken from wild animals. It is now true that farm land and ranch land hem in great lengths of the borders of many of the national parks. A few smaller game parks are already totally surrounded and in the next few years, probably no more than a decade, most of the big parks will also become islands in a sea of humanity.

In many parts of East and Central Africa farmers must contend with raiding elephants the way American farmers contend with raiding rabbits or crows except, of course, that in a single night one elephant can wipe out a season's efforts on an African farm. Small wonder that Africans shoot elephants when they raid their crops. The farmers do it themselves or the national game department rangers will do it on request.

Shoot them?! Don't those people realize that elephants are part of their natural heritage, a part of the wild heritage of every person? How can they be so ignorant and shortsighted? Words such as these are commonly exclaimed by American animal lovers. It is almost a cliché in America today to reply that that is exactly what our own forebears did when, as pioneers on the frontiers of a developing country called the United States, they virtually exterminated the bison, wolves, mountain lions, grizzly bears, and other wild animals that competed with farmers for land or threatened personal safety. That reply, though true, is not adequate. What did Daniel Boone know of ecology? Few frontiersmen worried about the "limits to growth." We today have the benefit of knowledge that didn't exist then.

The reason it is becoming so difficult to persuade most Africans to conserve wild animals runs a bit deeper. It is money, or, less succinctly, it is that conserving wild animals threatens one's well being and standard of living. Think back to your own vegetable garden or lawn or potted plant. After all the love and labor you put into tending it, some bug infestation comes along and kills your plants. Many a gardener who espouses wildlife conservation has, when threatened with tomato worms or aphids or some other marauding creature, slipped quietly down to the garden shop for a can of pesticide. Or, to pose the situation on a more drastic scale, suppose you have just discovered that your house is riddled with termites. What homeowner will, in response, open his eyes wide, fascinated by the remarkably sophisticated society and complex communication systems of termites? Some of the most distinguished biologists of our time have devoted their lives to studying the social structure and behavior of termites, ants, wasps, and other social insects. Their colonies are models of urban planning, with specialization of labor, administrative hierarchies, day care centers, armed forces, and more. Modern biologists regard the termite world as one of the pinnacles of social evolution. But even when confronted with all of this, it is difficult to imagine an American homeowner, even one with a college education and a love of nature, electing not to call the exterminator when termites threaten his house. The American's investment in his home is qualitatively identical with the African peasant's investment in his crops. And in Africa elephants may not only trample your fields, they may also push over your house or crush your children to death.

To poison the termite and condemn the African farmer who shoots marauding elephants is to display an immature attitude toward wildlife and the people who must live with it.

A sober and thorough examination of the elephant situation

in Africa has convinced many wildlife management experts and authorities on elephants that there is not the slightest possibility of preserving in the world anywhere near the number of elephants alive today. Much the same is true for practically every other large mammal whose habitat is in some developing country. Forest-dwelling species such as the leopard, gorilla, chimpanzee, okapi, bongo, giant forest hog, some elephants, and some buffalo are virtually certain to be reduced when trees are cut for wood or land cleared for agriculture. One of the greatest threats to the forests in developing countries is the need for wood as fuel. In much of Africa and Asia wood is the prime energy source for most people and forests are rapidly disappearing. Plains-dwelling species such as lion, cheetah, hyena, wild dog, jackal, giraffe, and most of the antelopes are certain to be reduced through the conversion of land either to crops or to cattle raising either by traditional, seminomadic herding or by modern fenced ranching. The Masai tribe, who are supposed to be particularly tradition-bound, are taking up Western-style ranching in Kenya and Tanzania.

Sometimes it is argued that such development is unwise for countries that depend heavily on income from tourism because it will eventually remove the things that attract visitors. Yet there is really little choice for the governments of developing countries and none at all for the people themselves. While tourism does earn substantial amounts of foreign exchange for some countries, few of the dollars or pounds or marks trickle down to the farmers and herders. Again, the example of the developing United States cannot be ignored. The development of this country was hastened dramatically by the federal government's allowing people to claim vast tracts of wilderness for homesteading, by opening the frontier with railroads, and by setting up the land-grant colleges which converted still more wilderness to direct human use. To farm their claims, the homesteaders first had to drive out or destroy most of the large wild animals living on the land. Many Americans have forgotten that the wheat and corn belts of the Midwest and West were once native grasslands as alive with wild animals as are the savannas of East Africa. They once held millions of bison, pronghorn antelopes, wolves, coyotes, prairie dogs, ferrets, and many other smaller mammals. A century or so ago the American heartland would have merited, as readily as does East Africa today, the distinction of being a "precious wildlife heritage." If anything, the vast herds of bison must have been more impressive than the Serengeti's herds of the much smaller wildebeest. If we had set aside a few thousand square miles back then and managed it properly, some part of Nebraska or South Dakota would today be the equal of the Serengeti or any other part of the world in offering a spectacle of wildlife abundance. But we didn't. We wiped out the American wilderness because we needed, or wanted, the land for something else. And, to a significant degree, we owe our present standard of living to the agricultural bounty of this once-wild region. Even today the grain and soybeans that these areas produce are this country's chief earner of foreign exchange.

Precisely the same phenomenon is occurring in South America, tropical Asia, and sub-Sahara Africa. Africa, again, offers the most dramatic example, for it has both the world's greatest abundance of wildlife and the world's fastest growing human populations.

Superficially it might seem as if Africa, whose 432 million people (roughly 10 per cent of the world's total) occupy 20 per cent of the world's land area, is comparatively under-populated. In fact, judged against the arable land available, Africa is already well over its carrying capacity. Only 10 per cent of Africa's land gets enough rain to support food crops. This gives Africans about half as many acres of productive land per person (1.2) as the United States (2.1). At present only a portion of Africa's arable land is under cultivation. The part that isn't planted to crops is heavily occupied by wildlife. To keep up with the demand for food, Africans are turning increasingly to these unexploited lands and even to less desirable lands where the rains are seldom plentiful or reliable. These include the grasslands that support the vast antelope herds that support the lions and hyenas.

Part of this agricultural expansion is made possible by the recent development of drought-resistant varieties of corn, a staple in African diets. Some varieties can reach maturity on as little as five to seven inches of rain in a season. In Kenya, which has only a third of an acre of arable land per person (one fourth the average for Africa as a whole), adoption of these new varieties is theoretically possible over most of the country. This would multiply the land-per-person ratio severalfold. At present, however, most of Kenya is sparsely inhabited by people and heavily populated by animals. But Kenya's 14 million people are expected to double their numbers in the next twenty years.

What lies ahead for African wildlife? Few who are familiar with the situation there doubt that the larger animals will be substantially reduced in numbers everywhere and totally exterminated in many areas. This has already happened in several African countries. Parts of West Africa, long the most heavily populated part of the continent, have been devoid of large numbers of wild animals for decades. In East, Central, and Southern Africa, where the wild species generally remain plentiful today, three countries are left with little more than remnant populations that have been herded inside parks. They are South Africa, Rhodesia, and Namibia (South-West Africa), all at this writing dominated by minority white rule and with long traditions of gun-happy hunters and settlers akin in regard for wildlife to the settlers of the American Middle West.

Many who are familiar with the African wildlife situation have already privately given up hope that any significant wildlife populations can be maintained outside the national parks or that the parks, which are rapidly losing the surrounding unprotected buffer zones, can sustain the numbers presently in them. They see an Africa ten or twenty years from now that will have as much spectacular wildlife in it as do, say, Virginia or Tennessee now. There will be gazelles grazing the roadsides at night as do white-tail deer in the United States. There will be squirrels and some birds. And that's about it. There won't be any more giraffes striding across the road or elephants coming to the river to drink or lions sleeping just beyond the hill. Those will be in protected parks just as the bears are in the Great Smoky Mountains. And, quite possibly, they will be so pestered by tourists that, far from behaving naturally, they will come up to the cars and beg for food. An elephant stretching its trunk to the car window for a hand-out might be a bit of a thrill but it wouldn't be an elephant behaving in the way that makes elephants so impressive and it would hardly be worth a trip to Africa when the nearest zoo could offer the same experience.

Even though the human populations in Africa stop at the park borders, the effects of proximity can easily alter the character of the parks in many ways. Many parks represent only a nuclear territory where animals spend most of their time, but out of which they must migrate at certain times of the year. The worst part of the dry season, for example, may force them to seek water or green vegetation outside the park for a few weeks. If that area outside the park is no longer available, the animal population will decline to the level that can be sustained entirely within the park during the least favorable season. People outside the park can also alter the lives of "protected" animals by diverting the water that flows into the parks or by triggering erosion that loads the streams with too much silt. Logging operations in the hills near Kenya's Tsavo National Park, for example, have caused so much erosion that the Voi River, which drains the slopes, has become silted up; it no longer flows during the dry season and animals are denied an important source of water. A third effect on park wildlife of surrounding human habitation is the compression. of herds of large animals that once were widespread into a relatively smaller area. Tsavo's elephant overcrowding is an example, even though it is one of the world's largest protected wilderness areas.

Whether or not the compression and reduction of wildlife populations will stop before species are pushed to extinction cannot be forecast with certainty. Africa has seen fewer extinctions in modern times than any of the other continents, but that does not mean it is impossible. Some of the more pessimistic authorities do not see how it will be possible to avoid many extinctions and wide exterminations.

If we cannot conserve all the species in the numbers existing today, which species should we try to keep large numbers of? How many of those species should we hope to retain? Or, for that matter, why should we keep them at all? These fundamental questions are only beginning to be addressed by professional conservationists and there is a wide variety of opinions on every point.

Most reasons given for the conservation of animals fall into one of three broad categories—ecological reasons, aesthetic reasons, and economic reasons. Each deserves some consideration both on its merits and on its value as an incentive for wildlife conservation. (An idea may, of course, have great theoretical merit but not persuade anybody who needs to be persuaded.)

The ecological arguments generally assert that all life on the earth is part of a complex web linking every form of life to every other and that if you destroy one part of the web, you weaken all the others. The earth supports man because of its biological richness; any diminution in this richness endangers the "life-support systems" that sustain man. There is certainly something to this argument. If we drained enough herbicide into the oceans to kill the algae, our major source of atmospheric oxygen, we might eventually suffocate. There are other examples to support the ecological argument, most of them involving organisms near the bottom of the food chains. By contrast, there is little evidence that the survival of the whooping crane, for example, or the Sumatran rhinoceros or even the African elephant is essential to human survival. In fact, most of the animals that arouse popular interest in conservation are near the tops of food chains and well outside any human life-support systems. In some instances the decline of a large animal can be an indicator of more fundamental problems. If peregrine falcons cannot reproduce because of DDT in the environment, this can be a tip-off that DDT may be harmful to us as well. But if peregrine falcons were to disappear, that would not be of major ecological impact in and of itself.

If the extinction of the golden lion marmoset of Brazil or the mountain gorilla of Rwanda or the Siberian tiger diminishes the healthiness of our environment, which is open to question, the impact has already been felt, for these animals are nearly gone already. Their present population, like that of so many other species, is but a small fraction of what it once was and the damage, if any, is done. If you really want to worry about ecological damage to the planet, look closer to the ground—at the worms that keep the soil fertile and aerated, at the insects that consume dead and decaying matter, at the bacteria that carry out so many of the chemical reactions necessary to higher forms of life. Although official conservation bodies are just getting around to putting insects and other invertebrates on the endangered species list, there is still no appreciable constituency for these creatures, no one to picket on behalf of a snail or an ant. It will take a lot of conservation consciousness raising to get a Save-the-Clostridia Foundation, clostridia being a major group of soil bacteria.

The ecological argument *is* valid when offered in defense of whole ecosystems, such as salt marshes—the coastal shallows that are an essential hatchery and nursery for the young of so many fish and shellfish. Destruction of too much wetland could spell the end of many commercial fisheries and, in fact, substantially reduce the supply of food to larger marine animals such as sharks, toothed whales, seals, and others. When the ecological argument is offered in defense of the leopard or the gorilla or any of the other large land species so favored by citizen conservationists, however, it is a weak one.

The second category of reasons, or incentives, for saving large animals is aesthetic. Sometimes these are stated in rather personal terms: "The animals are beautiful to look at," or, "Even though I may never get to Africa, I just want to know they are there," or, "I don't want my children to grow up in a world with no lions or tigers." By "aesthetic" I do not mean sentimental imaginings about animals but a mature appreciation of animals for what they really are. I include in this category the educational value of animals. This includes more than the scientific study of some species that may shed light on basic questions affecting mankind or yield animal models for disease research or provide some other practical benefit. If it did not include more, then once that knowledge were gained, the animal might then be considered dispensable. Beyond the scientific worth of a species is its educational value, or "fascination value," to nonscientists, which is something that each person gains for himself. It thus demands a continuing availability of the animal so that each new generation of people might reap that benefit. No matter how good wildlife films might be, or books, or zoos, these are not fully satisfying substitutes for the aesthetic pleasure of observing the living animal in its natural habitat. Even if most of us never get to Africa or the Amazon rain forest, just knowing that it might be possible someday will be a powerful incentive to learn about wildlife.

There is, however, a larger dimension to the aesthetic argument. It is that intangible quality that people seek when they try to "get back to nature" or to "commune with nature." Writers have described the psychological value of this for centuries. Henry David Thoreau is the best known American example. If we destroy nature, then we won't have much to commune with, or if we remove from an ecosystem its largest animals, we will have only a part of nature with which to commune. Even if our communing is limited to Central Park or to a quiet country stream just out of town, much of the value we derive from it is based on the knowledge, as the meditating mind wanders, that there are other such tranquil, mysterious natural places in the world where giraffes amble or wolves prowl. If those truly wild places are no longer there, then Central Park or that country stream become not the psychological gateway to a wilder world but dead ends.

The aesthetic argument, then, is an intangible one, but it is, for many people, a powerful and persuasive one. But there is a practical problem here. Aesthetic values are shaped by one's culture. They are, as anthropologists say, "culture-bound." You may thrill to the sight of a lion stalking its prey in the Serengeti. A Masai tribesman, who loses his cattle to lions and who must carry a spear to protect himself and his herds, would feel differently about lions.

The enthusiasm which many Americans and other Westerners show toward the wilderness and its animals is not shared by peoples everywhere. In fact, to a Masai or any of hundreds of other ethnic groups who live their lives close to nature, there is no such concept as wilderness. What seems wild to us is tame to them. This is illustrated for North America in the comments, from the previous century, of Chief Luther Standing Bear of the Oglala Sioux: "We did not think of the great open plains, the beautiful rolling hills and the winding streams with their tangled growth as 'wild.' Only to the white man was nature a 'wilderness' and only to him was the land 'infested' with 'wild' animals and 'savage' people. To us it was tame." Wilderness becomes recognizable as such only in contrast to what we call "civilization." It's the old story of "you don't know what you've got until you lose it." Americans destroyed most of their own wilderness before they realized its value, and now they have to work to experience wildernesses elsewhere.

A century and a half ago, when there was still a lot of real wilderness to be seen in eastern North America, it went largely unrecognized for its aesthetic worth by Americans living on the "frontier." It took a Frenchman, Alexis de Tocqueville, whose forebears had lost the wilderness of France centuries earlier and who came to the underdeveloped America of the 1830s, to appreciate it. After him a long succession of rich Europeans came to America, got themselves outfitted for the woods in local shops, and hired native guides to take them into the wilderness to hunt or simply to look. This provided no end of amusement for many Americans, who could see only a lot of rugged wasteland to be tamed and little in a bear to admire.

Now that most Americans as well as Europeans are city folk, the attitudes have become very different. We want what remains of the wild animals and their wilderness to stay the way we like to think of them. And now that the demand for wilderness is so large and the supply so small, the insistence of rich Western countries that poor countries set aside their wildernesses for Western benefit is becoming increasingly strident. By the same token, the need of the poor countries, whose cultures give them a different sense of aesthetics, to use that land for their own purposes is growing stronger.

Consider what the wilderness demands of the rich countries, defined by their own culture, has meant to many Africans. At the most primitive end of the African technological spectrum are the people who have always made their living by hunting or by seminomadic grazing of their herds of cattle and goats in what amounts to open range country. Their way of life has continued happily for centuries without putting too much pressure on the environment. In fact, studies of several tribes reveal that they have an intimate knowledge of how to adjust their hunting or grazing patterns to avoid overtaxing their environment. These people have never thought of themselves as poor or primitive, and in terms of human relations and social justice they are indeed rich. Along comes their new national government, striving toward Western ideals and saying that henceforth it will be a crime to hunt in such-and-such an area and it will be illegal even to let your cattle graze there. Why? Because rich foreigners want to come to the area and see the animals or even to shoot them. The African hunter or herdsman watches incredulously as total strangers drive their vehicles across the plains, killing more grass than his cattle would eat, or as foreigners shoot the game he is forbidden to kill. It is okay for these alien people to indulge their pleasures but a crime for the African to earn his living on what had been his land. Herders who cannot believe the existence of such laws or who prefer their own traditions have had their cattle shot by park rangers. Hunters who continued to gather needed protein for their families have been put in jail as poachers. It is not easy to convince these people that they have the wrong sense of aesthetics about the "precious natural heritage."

Further up the technological scale are the Africans who make their living by cultivating the land and the growing millions in the cities and towns who depend on the farmers for food. A father's land can be divided among his sons only for a few generations before there is not enough for any one son to make a living. The children must seek land elsewhere or give up agriculture and go to the city, find a job, and further increase the food needs of the landless city dwellers.

In the face of this Western pressure to conserve wildlife, some of the world's poorest countries have made heroic efforts. Tanzania, for example, gained independence from the British in 1961 with just one national park, the Serengeti.

Since then it has created nine more, most of them large by American standards, and the country plans to create still others, locking away more and more land from its own people. Keep in mind that African parks are not just deserts or mountains or other scenic but desolate areas. They are rich and diverse ecological zones, abounding in protein on the hoof and the vegetation to support it. Zambia, following the same course, has now put one third of its entire area under protected status. Zaïre is working toward a goal of putting 13 per cent of its land in national parks. The comparable figure for the United States is 2 per cent of its land area. Although a small conservation constituency is developing within African countries, largely among the urban and educated classes, further efforts to set aside more wilderness are encountering growing popular resentment for obvious reasons. Someday, when Africa's economic development is further along, the aesthetic argument for conserving wildlife may be accepted there, but for now it does not carry much weight with the majority of Africans. And their opinions count.

This brings us to the third set of reasons for conserving wildlife—economic. These, in the view of a great many experts on wildlife conservation, are the only incentives that have any promise of being accepted widely in the developing countries. The economic reason can be put in very few words —the conservation of animals must be made more economically attractive than any alternative use of the animals or the wild land. Put another way, wild animals must pay for themselves. If they cannot, some method has to be found whereby the wealthy, either as individuals or as countries, are willing to underwrite wildlife. Assigning a monetary value to a giraffe or an impala may seem crass, but it is not a great deal different from the manner in which wealthy people already support art museums or ballet companies.

There are two main ways in which wild animals can be made to produce revenue. One that is already widely accepted is through tourism, a leading earner of foreign exchange for several African countries rich in wildlife. Westerners come to Kenya, for example, to see wildlife and, in the process, spend considerable sums on hotels, safari guides, meals, vehicle rentals, and the like. This is all well and good as far as it goes, but there are some problems. Most of the income goes to wealthy entrepreneurs and does not trickle down to the peasants who suffer most by being dispossessed from the protected lands. The fact that tourism is a leading earner of foreign exchange says more about how small the other earners are than it does about tourism. Although Africans and Westerners who live in Africa think that the countries are overrun with hordes of tourists, the actual numbers are minuscule by American standards. It is a busy day in most African game parks if a hundred vehicles arrive. More people go to Coney Island on a summer afternoon than visit Kenya in a month.

The other source of revenue from wildlife is presently much less lucrative but, in the view of many observers, could be vastly more so. It is the harvesting and selling of animal products, from meat for local consumption to hides and ivory for export.

An example is leopard skins for which there is a sizable demand. Although the leopard is legally protected in Africa and may be taken only in very limited numbers, hunters shoot several thousand a year to supply the overseas demand. Because leopard hunting on this scale is illegal, all the money from these sales, running well into the millions of dollars, goes into criminal hands and is lost to wildlife conservation and to the African people. Advocates of controlled cropping argue that if there were a legal leopard hunting industry, the numbers of animals taken could be reduced through quotas and a hefty tax on each skin could go directly to benefit conservation programs. Part of the revenue would also pay for a policing staff capable of regulating the hunting far better than the poorly staffed game and customs departments of African countries do now.

Although it is popularly thought that the leopard is so reduced in Africa that any hunting at all threatens the survival of the species, Norman Myers, a Britisher and a recognized authority on African wildlife and a consultant to the private International Union for the Conservation of Nature, estimates that there are at least one hundred thousand leopards in Africa and possibly several times as many—some authorities estimate there are a million leopards. Myers figures that an annual off-take of ten thousand skins could easily be sustained, given the relatively high reproductive potential of the leopard. "If the trade were strictly regulated and the profits more equitably distributed," Myers says, "the African stockman would derive benefit from the creature which he at present looks upon with the same spirit as the American rancher views the coyote and the cougar." Myers observes that diamonds are another African resource desired by the wealthy and that the profits from the diamond industry have been used effectively to police the harvesting of that resource. Several species of leopard live in Asia but statistics on these are too scanty to calculate any economic potential.

Many other potentially revenue-producing animal species are known, such as elephants for their ivory, and even zebras, for their skins, which currently fetch $300 apiece in the United States. Preservationist sentiment, however, has made the sale of zebra skins illegal in New York and California, heretofore the two largest markets for the hides. These well intentioned laws effectively removed an incentive for the conservation of zebras as a renewable resource. When a zebra skin could bring an African rancher more than the meat from a cow, the rancher would have had good reason to ensure a continuing supply of zebras, since he would derive direct benefit from the sales of their skins; otherwise he would just as soon eliminate the zebras because they compete with his cattle for grass.

Because enlargement of the economic incentive means killing a certain number of animals, many wildlife enthusiasts are appalled by such suggestions. Their attachment to an idea of a pristine and inviolable wilderness prevents their accepting any reason for killing an animal, or at least for killing a "good" animal such as a zebra or an elephant or a leopard. A hyena or a shark might be okay to kill and certainly man's domestic animals all the time, but not one of God's good wild creatures. As many people have noticed, such attitudes are most common in cities, less common among those who live in the country, and least common among people who spend much or all of their time close to nature. In fact, the strength of the anticropping sentiment among wildlife enthusiasts is so strong that a number of international conservation groups that have debated the pros and the cons of controlled cropping have tried very hard to keep their deliberations quiet for fear that they would lose the support of their lay members.

Again the African leopard is a case in point. Until the leopard was placed on the official United States list of endangered species in 1972, thus prohibiting importation of leopard skins or trophies, there was a long, loud and emotional campaign to discourage the purchase of leopard-skin coats. Furriers were picketed. Prominent women renounced their leopard-skin coats publicly and vowed to wear only fake furs. Millions of people were led to believe the leopard was on the brink of doom.

Over the last few years, however, detailed studies of leopards have led to the new conclusion that the species is, in fact, still well established in almost every African country south of the Sahara and even plentiful in several. The very fact that thousands of skins were taken every year testified to the abundance of the species. An endangered species could not exist in the numbers to support the rate of hunting then going on. When news of the leopard's abundance came out, one might have expected the animal's devotees to rejoice. Instead, they declared the new reports to be false and refused to accept the news. Organizations and experts that had been about to suggest renewed hunting of leopards to produce revenue that could support conservation of other species backed off. The public, it appeared, did not want to believe that their beloved leopards were abundant.

If wild animals are not harvested on a sustained-yield basis to produce the revenue needed to justify their survival, there is one other economic stratagem. Those who wish to see the animals preserved must find a way to channel more money from the wealthy West to the impoverished countries where the cost of conservation is now borne. After all, it would be hypocritical of the West to say that wildlife is the "common heritage of mankind" and then refuse to bear a share of the burden. Myers, who has devoted much of his thought to this issue, suggests that the wild populations be designated a global resource under some kind of international agreement that would require all countries to contribute to their conservation. "In other words," he says, "the world community would somehow have to channel a cash subvention to the African countries in question year by year, in order to compensate their citizens for the 'opportunity costs' they experience in not putting the wildlands to 'more appropriate' purposes."

There is not much precedent for this kind of thing. It involves some of the same questions that have proved so difficult for the world's nations in developing a "law of the sea" and a means of compensating the victims of, for example, air pollution that drifts across national borders or water pollution in boundary waterways. Proposals similar to Myers' were discussed at the Stockholm Conference on the Human Environment in 1972 but they have not been moved substantially toward enactment. In principle these proposals resemble those already in operation in developed countries whereby tax write-offs and other economic incentives are granted industries that install and operate pollution-control equipment.

Not all the destruction of wildlife and its habitat in the developing countries today is due to the actions of the people of those countries. Some of it is caused by large corporations of the developed countries that move operations into the poor countries, for example, to clear tropical forests for lumber or to establish huge beef ranches. The forests of Southeast Asia and the Amazon Basin of South America are particularly vulnerable to Western and Japanese timber companies that clear-cut thousands of square miles of tropical forests every year. It is estimated, for example, that in ten years all the accessible virgin forests of West Malaysia and the Philippines will be cut down. More than five hundred timber companies have contracts to cut down Indonesia's forests. Volkswagen's Brazilian subsidiary is planning to turn a wilderness area in Brazil into a four-thousand-square-mile cattle ranch to supply the beef demands of the more developed countries. Much of the hamburger used by American fast-food restaurant chains comes from Central American ranches that, before the advent of these chains, were virgin tropical forest.

The forests at issue in these instances are among the least well-known ecological zones in the world. It is believed they harbor thousands of species of animals unknown to man and tens of thousands of plant species. If those forests are truly a part of the wilderness heritage of Americans, then Americans should be in a good position to help protect them by discouraging American industries from carrying out their share of the planned exploitation. This avenue should be easier to travel than those involving complex international agreements.

But still, the situation is not so simple. Consider the clearing of tropical forests to create rangeland for cattle. Given the long-term demands for food in the world, there is strong pressure to stop using so much grain to feed American cattle. Since there is a growing world-wide need for the high-quality protein in beef, many people see increased range-feeding of cattle in the tropical countries as one answer. Cleared tropical forests are generally not good for growing grain for human food. The soil is too thin. The soil will, however, often support grass for cattle. The idea is that if the beef demands can be met from these newly created pastures, more of the grain potential of the rich American soil can go directly to meet the growing need for grain to feed people. Thus, to argue that all of the tropical forests should be preserved as wilderness is to argue that the world should not produce all the food it can. To argue that the African herder and cultivator should not expand into the wild lands of his own country is to pressure African nations into continued dependency on costly outside sources of food, sources that cannot forever meet the world's escalating needs. Clearly, to keep *all* the present wilderness would be to condemn even more millions of people to chronic malnutrition and, in times of drought, to even wider famines than the world has already seen.

Of course, the root problem is the population explosion. Of course, the underdeveloped countries should cut their birth rates. Of course, the world's population cannot continue growing forever. But there is no getting around the fact that even if every country in the world were to achieve the two-child family overnight, the world's population would still continue to grow for several decades. The United States birth rate is already slightly below the so-called replacement rate—two children to replace two parents—but the American population will not stop growing until well into the next century because more potential parents have already been born than there are current parents. There is absolutely no doubt that the world's population will continue to grow for several decades, even, to repeat, if the two-child family were to become the rule immediately—and that is not going to happen. We have a choice either to give up much of the wildlife and the wild habitat and fall back to wilderness enclaves or to condemn huge sectors of mankind to famine. The only questions for us if we want a humane future are where we should establish the enclaves, how to draw the boundaries so that the protected domains will remain viable as wilderness areas, and what wildlife management techniques to employ to offset the inevitable deleterious effects of erecting artificial boundaries around natural areas. It is an enclave strategy, and to many people that will seem not much better than a zoo. But, with very few exceptions, I do not see how we can reasonably demand more in a humane world.

Assuming that there are valid reasons for conserving wildlife in a natural setting, or at least in as natural a setting as possible, then on what scale exactly should we attempt to protect wild animals and their habitats? The quick answer, of course, is "on as large a scale as possible" and, until now, that has been the one-sided viewpoint of most conservationists. But, as the preceding discussion has labored to point out, that view is selfish and immature.

Obviously there is no one answer. Some animal species require very large regions in which to live; others need only small regions. Some animals are more popular with tourists and would need to be available in many locations; others are of little interest to anyone but biologists and would need to be protected in only a few places. Some species are easy to capture, transplant, and establish in other areas; some species cannot be transplanted so well and must be protected where they are now. Some ecosystems are so complex that the ones nature has taken centuries to establish must be protected; others are relatively simple and can be created by ecological engineers in lands presently used for other purposes. Some people will be satisfied only by a pure wilderness with no roads or restaurants or boardwalk viewing areas; other people would not be interested in or capable of visiting wildlife areas without such amenities.

It may be helpful at this point to examine the various scales on which wild animals may be conserved. At one end of the scale is the zoo where about the only thing protected is a tiny population of the animals themselves. Whether it is humane to keep animals caged in zoos is an issue so well discussed elsewhere that it will not be dealt with at length here. Suffice it to say that there are very few wild species that behave naturally in zoos. Food-seeking behavior, for example, is the main preoccupation of most wild animals, but in zoos there is noth-

ing for animals to do but pass the time until the keeper brings
their food around. A bear eating wild blackberries or fishing
in a mountain stream is far more bearlike than one rearing up
on its hind legs in a zoo, begging for popcorn. Most zoo enclosures are so small the animals never have enough space in
which to run. Zoo-goers have no idea that a crocodile can pace
a man running or that a loping wolf actually seems to float, its
feet almost appearing not to touch the ground. The number of
any large species kept in a zoo seldom approximates the size
of natural social groupings. However striking one elephant
may be, it is nowhere near the spectacle of the families of
fifteen or twenty or, occasionally, several hundred to be seen
in the wild. And, of course, the setting in which zoo animals
are displayed rarely approximates their natural habitat. Never
will the zoo-goer see a large predator preying.

In some respects a much better appreciation of wild animals can be gained from natural history museums, with their
stuffed animals set in lifelike poses and displayed in authentic
reproductions of their natural habitat. To be sure, there is
something exciting about seeing the animal—breathing, mus-
cles rippling as it walks—but it detracts from the effect to
know that what the animal is doing is rarely what its free-living kin are doing. One might even argue that an animal in a
zoo is not a whole animal for it is lacking its natural behavior.
Every animal's existence is more than just the flesh and blood
contained within its skin; it is this, plus its natural environment, the two linked in a dynamic equilibrium. A person with
a mature appreciation of wild animals recognizes this linkage;
unfortunately he or she is also disappointed or even saddened
by its absence.

Nonetheless, for those species that are able to breed in captivity, but which are threatened or already extinct in the wild,

zoos are necessary for preserving the genes of the species. The zoo becomes a gene bank, keeping the spark of a species alive until there is some opportunity to re-establish a free-living population. This is already the case with several species that are extinct outside of zoos such as Przewalski's horse. The European bison, also called the wisent or aurochs, once was confined entirely to zoos until a breeding colony of captive animals was put out into a protected area in Europe.

More natural than zoos are the free-living populations of remnant species that exist entirely on small patches of protected land each measuring only a few dozen square miles. These species have been exterminated from nearly all (usually more than 99 per cent), of their former ranges and now survive by the skin of their teeth through intensive management programs in one or more tightly guarded areas. Examples are the whooping crane, the Sumatran rhino and the Bengal tiger. The whooping crane, though it has not numbered more than seventy animals in decades, is regarded today as fairly well protected. The advantage of this method over zoos is that the animals live much more natural lives even though, for example, human beings have to control the water level in their nesting grounds or fence off the habitat. Conservation workers must take steps to compensate for all the large-scale environmental factors that affect a natural habitat but which would not be likely to operate in a small area. An example of these large-scale factors is water: a river upon which the animals depend for drinking might flow into the animals' area from, say, agricultural land where it could be tainted with silt, fertilizer runoff, pesticides, and the like. Artificial control measures must be used to protect the animals. Many people find this arrangement more aesthetically satisfying than zoos. Also, there are many species that do not breed well in captivity and this living arrangement for them is the only alternative to extinction. A major disadvantage of such conservation areas is that they are usually few, remote, and unavailable to large-scale public viewing. Most animal lovers will just have to settle for "knowing they are there." Another disadvantage is that normal predator-prey relations are not always possible. It is easy for most people to see that a predator such as a lion or a wolf is not living a truly natural life if it is not allowed to hunt and kill its own food. What most of us are going to have to realize, however, is that it is just as unnatural for a prey species such as deer or zebra to live without being preyed upon. Blindness to this fact has prompted many sentimentally inclined conservationists to argue that where deer are living in a small area without natural predators, human hunters should not be allowed to prey upon the deer. These sentiments have led to laws preventing such hunting. Consequently the deer populations in such places are growing rapidly and outstripping their available food supplies. A better approach to such situations would be either to reintroduce wolves or to permit human hunters to remove the excess deer populations. Man, after all, has been a part of many animals' habitats for millions of years, preying upon such animals as naturally as lions or wolves. The deer in North America have been hunted by people for many thousands of years and it should seem unnatural to outlaw hunting that would restore, not destroy, a natural balance.

A still-larger scale of wild animal conservation is the typical big national park or wildlife refuge where relatively large populations of animals are allowed to coexist under the natural system of checks and balances. Of course, "national park" is an administrative designation implying the governmental level at which protection is enforced. The term is used here to mean a very large conservation area on the scale of Yellowstone or the Serengeti, regardless of the governmental unit administering it. Predators are part of these systems and are allowed to eat their fill. With the larger areas, often hundreds or thousands of square miles, management programs seldom need to be elaborate; there is enough variation in the ecological environment for animals to escape adverse conditions such as fire in one region without having to leave the park. With the large area it is also possible to permit a sizable number of visitors without disrupting animal behavior patterns. The behavior of the animals in such areas is usually entirely normal and natural. By most criteria this is the most desirable scale of conservation, but it also poses enormous disadvantages in that a large area of land must be put out of bounds to other uses such as agriculture or, to cite conflicts common in the United States, energy-intensive recreation such as powerboating, snowmobiling, using all-terrain vehicles, and the like, or mineral or lumber exploitation. Even establishing a woodland picnic ground can be incompatible with wildlife conservation, given the slovenly and noisy habits of many picnickers. Land costs money and, if it is not already in government hands, costs of further acquisition by the government may be prohibitive. The national park or its equivalent in size is where, for most of the world, the formal conservation of animal species and their habitats is practiced on its largest scale. There are, of course, laws that make it illegal to kill a given species anywhere, but, as noted earlier, hunting is a minor threat to most species compared with the destruction of their habitat.

Readers who are familiar with wildlife conservation programs will know that this three-level series of conservation methods—zoos, small and intensively managed remnant areas, and large national parks—is oversimplified. There are in-between possibilities and strategies of multiple land use. It is not the intention of this book to lay out all the possibilities but to illustrate a way of thinking about animal conservation, to show that there are alternatives, with various costs and benefits.

For example, if we wish to have cheetahs in the world with us, we must decide how much it is worth to us to have them. Cheetahs are rapidly disappearing in Africa. They are much closer to extinction than leopards or lions. It is estimated that there are fewer than twenty-five thousand left and possibly fewer than ten thousand. The number, whatever it is, is believed to be half what it was just fifteen years ago, and experts project the present rate of decline to leave the world with only five thousand to twelve thousand by the end of the next decade. By comparison, there are anywhere from one hundred thousand to one million leopards in Africa.

Cheetahs live in open country and hunt by day. Thus, unlike the secretive and night-hunting leopards, they are very conspicuous. African ranchers consider them pests and farmers want to clear their habitat for plowing. Hence, within one generation there will be very few, perhaps no cheetahs in Africa living outside protected areas. What scale of conservation should be undertaken for them?

Cheetahs breed well in captivity and could presumably survive indefinitely in zoos, though they would probably never be allowed to run down and kill their own food there. The second alternative, the small protected area, is also possible for them, but it would provide room enough for only very few individuals. Also gazelles, the cheetah's prey, must be able to have a continual supply of short grass; they will not eat long grass. Under completely natural conditions, the gazelles are nomadic, moving from one place where they have eaten up the proper grass to another place where, following a rain, tender new grass has just sprouted. In a small protected area, the distribution of rainfall might not assure a supply of the preferred grass. For all these reasons, we can see that a conservation area on the scale of a large national park is the only possibility for sizable populations of cheetahs. This is costly in terms of compensating people in the immediate area for their loss of opportunity to use the land for something else. Those who wish cheetahs conserved in this fashion—if they would do so with a meaningful regard for competing human needs—would have to be willing to compensate Africans living nearby. The exact mechanism of compensation is beyond the scope of this book, but it could take any of several forms, from direct cash payments to Africans, to helping them set themselves up in tourist businesses, to establishing a cheetah hunting or harvesting program from which the proceeds would go to the Africans.

Even if one such cheetah conservation area were equitably established, the question would remain as to how many were needed or wanted. If the compensatory system is simply a direct payment of money (derived from, say, an international wildlife tax or some system like that used to support the United Nations), there would be an obvious limit to the volume of cheetah conserving, since it would depend on the budget of some hypothetical international wildlife authority. However, if the compensatory system is based on the proceeds from cheetah marketing (to tourists, sport hunters, or furriers), then as many cheetah reserves could be established as the market would bear. If there really are a lot of people who want to see cheetahs in the wild, the fees collected might be enough to support several cheetah sanctuaries. If not, then one must decide why we need to conserve so many cheetahs.

Cheetahs have been discussed above as if one were setting out to conserve only cheetahs. This simplifies thinking about the concepts involved, but it should be kept in mind that cheetah habitats include a wide variety of other species, plant and animal, and that all would benefit from the conservation effort. A single large-scale conservation area such as the Serengeti National Park, for example, contains within its six thousand square miles not only 150 cheetahs but also two thousand lions, three thousand hyenas, upwards of five hundred thousand zebras and wildebeests, half a million gazelles, and a hundred thousand other large animals such as buffaloes, elands, and giraffes.

How many cheetahs or lions does the world need? Does it take more than half a million gazelles to assure animal lovers that "they are there"? Perhaps the most crucial question is whether the Serengeti is big enough to handle the world tourist demand for the animal species represented in the park. If there were no other region in the world to serve as a lion or cheetah viewing place, the Serengeti could not handle the demand. For that reason alone, one could argue that the world needs several such places. I suspect the number is not very much larger, however. With better means of traveling through the park than in hordes of vehicles wandering all over the fragile grasslands and other efficiencies, the world might need no more than half a dozen Serengetis. As of now, Africa could easily supply this number and turn a number of parks back to the people of Africa.

There are other types of habitats in the world that one would want to preserve too—rain forests, deserts, mountain forests, swamps, and so on. Some have less appeal to the public and therefore one could argue that fewer would need to be set aside. At a minimum, the world probably needs to keep at least one large example of every major type of ecosystem to be found. We should ensure that nothing will harm at least one area of, for example, an Amazon rain forest—at least ten thousand square miles of it to be kept as it is, so that, at a minimum, scientists may study it for whatever it has to teach us. We should keep similarly large tracts of Central Asian steppe, West African Sahel, North American prairie, and every one of the several dozen ecologically distinct zones, or biomes, around the world. This is, of course, not a new idea. Ecologists and others have long argued for such a plan and several such areas are already under protection. But more need to be added.

One significant new program that approaches these goals is already underway. It is called the "biosphere reserves program" and it aims to designate a number of ecological zones that are to be studied scientifically with an eye to establishing detailed programs for their protection. This effort is jointly sponsored by the United Nations Educational, Scientific and Cultural Organization (UNESCO), in co-operation with the National Science Foundation in the United States and a similar agency in the Soviet Union. Some thirty biospheres, or ecological zones, within the United States have already been designated for study and possible permanent inclusion in the program. Many of the areas are already protected as national parks and wilderness preserves.

Whatever the ultimate system of protection, most ecologists feel that the more fragile areas ought not to be heavily visited by tourists, for any sizable influx of people could disrupt the natural scheme of things—assuming, of course, that native people are not already fully natural parts of the areas' ecosystems, as they are, for example, in parts of the Arctic and the Amazon basin. An ecological area that was of great popular interest—like the Serengeti, for example—would have to be divided into several smaller parks, with an "open door" policy for tourists in most, more restricted access enforced in a few, and one park off limits to all but bona fide scientists. A system of tests could even be devised for people to take—covering ecological awareness, camping skills, etc.—in order for them to become certified to enter the various subareas. People who pass the strictest tests would be allowed into the most fragile areas; those who qualified at a less sophisticated level would be allowed into hardier ecological zones than the first group; and the majority of tourists would qualify for the hardiest area of all. Such licensing systems are already used in some countries to control the access of hunters to wild areas. Some wilderness areas of the United States already do this to some extent by outlawing vehicles. Licensing systems are used almost universally to limit access to certain professions where incompetence can be disastrous and even to restrict the types of cars or trucks people may drive on the road. It may be time that fragile wilderness areas enjoyed similar protection.

How can all the complex and even authoritarian-sounding proposals made in the foregoing discussion possibly be made to work in today's world? The answer, I suggest, may lie in the establishment of something like an international wildlife conservation agency, ideally under the United Nations and supported by the UN member nations under the usual cost-sharing arrangements. Such an organization, with a scientific staff for ecological advice, could establish global conservation priorities and promulgate standards for protection of wildlands. Most importantly, it could be the channel through which money from the wealthy countries is given to the poor countries to pay the price of conservation. No country is likely to grant an international agency direct rights over parts of its own territory, but many countries might well organize their domestic conservation programs in accord with international standards and put up some "earnest money" themselves if they expected to receive an amount of money from the international agency that was large enough to make conservation worthwhile.

The same agency could also help control any trade in wildlife products by establishing their legality under international law and assisting in regulation of that trade to ensure that it did not exceed ecologically justifiable levels and that the profits went to the proper people.

Although the financial arrangements of such a system might seem totally unworkable, there is ample precedent in such organizations as the World Bank and the various regional development banks for making money available to poor countries on attractive terms to carry out various agricultural and industrial development programs. Low-interest loans from international banks to poor countries might, for example, be made to help these countries establish conservation areas and the facilities necessary to make the conservation area pay. The terms of repaying the loan could be indexed to the rate of earnings from the conservation program. The amount of the loan would go not only for the immediate costs of conservation but to finance alternatives for lost economic opportunities. For example, if creating a conservation area would deny herders a place to graze or water their cattle, the loan should cover the cost to a nation of establishing an alternative water source and irrigated land outside the park that would produce enough grass for the herders' cattle.

This is, of course, not a detailed proposal for a new international agency. It is merely a suggestion or an indication of a new way for conservationists to start thinking. People who want to keep wildlife in the world, for whatever reason, are going to have to find a way—maybe not this one, but some way—to put up or shut up.

The time is approaching to put away our animal myths and immature conceptions of what the earth really is. There may seem to be perfectly good reasons to want to preserve every species alive, but the fact is that the extinction of a species is as natural as the death of an individual. Without extinctions there would have been no evolution once the world was filled with animals; and with no evolution there would have been no *Homo sapiens*. The glory of the world is not that everything was created just so, but that everything changes. And in that world man is not an alien intruder.

There is room on our planet for lions and gorillas and whales but we must understand them as they truly are and not belittle our own species by forgetting that we too are a part of nature.

Index

About the Author

Boyce Rensberger is a leading science writer at the New York *Times* and a contributor to the New York *Times Magazine*. Formerly he was a science writer at the Detroit *Free Press*, where the managing editor called him "a rarity . . . the science writer whose writing is as good as his expertise." His writings on ecology won the Scripps-Howard-Meeman Foundation award for conservation writing in 1970. In 1973–74 he was awarded an Alicia Patterson Foundation fellowship both to do research on the origin and evolution of man at the major digs in Africa and to examine wildlife conservation practices there. Mr. Rensberger studied biology and anthropology as an undergraduate at the University of Miami and as a graduate student at Syracuse University.